A. Tognoli (Ed.)

Singularities of Analytic Spaces

Lectures given at a Summer School of the
Centro Internazionale Matematico Estivo (C.I.M.E.),
held in Bressanone (Bolzano), Italy,
June 16-25, 1974

C.I.M.E. Foundation
c/o Dipartimento di Matematica "U. Dini"
Viale margagni n. 67/a
50134 Firenze
Italy
cime@math.unifi.it

ISBN 978-3-642-10942-3 e-ISBN: 978-3-642-10944-7
DOI:10.1007/978-3-642-10944-7
Springer Heidelberg Dordrecht London New York

Reprint of the 1st ed. C.I.M.E., Ed. Cremonese, Roma 1975
With kind permission of C.I.M.E.

Printed on acid-free paper

Springer.com

CENTRO INTERNAZIONALE MATEMATICO ESTIVO

(C.I.M.E.)

2° Ciclo - Bressanone dal 16 al 25 giugno 1974

SINGULARITIES OF ANALYTIC SPACES

Coordinatore: Prof. A. TOGNOLI

CENTRO INTERNAZIONALE MATEMATICO ESTIVO
(C. I. M. E.)

H. HIRONAKA

SPECIAL CLASSES OF SINGULARITIES OF CURVES AND SURFACES

(The text was not delivered by the Author)

Corso tenuto a Bressanone dal 16 al 25 giugno 1974

CENTRO INTERNAZIONALE MATEMATICO ESTIVO
(C.I.M.E.)

F. LAZZERI

ANALYTIC SINGULARITIES

Corso tenuto a Bressanone dal 16 al 25 giugno 1975

F. Lazzeri

Introduction.

1. Let X be a "space" and $(X_t)_{t\in T}$ a "continuous family of subspaces of X" ; in general that means that one has a "total space" $\tilde{X}$, a "moduli (or parameter) space" T and morphisms $\tilde{X} \xrightarrow{\pi} T$, $\tilde{X} \xrightarrow{\varphi} X$ where the family $(X_t)_{t\in T}$ is the family of the fibres $(\pi^{-1}(t))_{t\in T}$ of π and for each $t \in T$, the restriction of φ to X_t is an embedding of X_t in X . Obviously the meaning of the words subspace, fibre, embedding has to be specified depending on the geometric context (algebraic or analytic geometry, differential topology etc...) in which one is working. What happens in general is that there exists a closed subset Δ of T (the "discriminant locus") s.th. locally on $T - \Delta$, π can be viewed as the projection map of a product space; in particular on each connected component of $T - \Delta$, the fibres X_t are equivalent to each other.

Similarly one can consider the continuous family $(X_t)_{t\in\Delta}$ (instead of $(X_t)_{t\in T}$) and one finds as before a closed subset Γ of Δ ; by repeating this process one tries to stratify the map π and to classify (i.e.. to describe) the fibres of π , i.e. the elements of the continuous family .

Alternatively instead of attempting to stratify the family as above, one can examine the family $(X_t)_{t \in T-\Delta}$ more closely. Fix $t_o \in T - \Delta$ and associate to each loop γ in $T - \Delta$ with base point t_o a continuous family $(X_s)_{s \in S^1}$ parametrized by the circle S^1 ; by associating to such a family some (in general topological) invariant that depends only on the element represented by γ in $\pi_1 (T - \Delta, t_o)$ one obtains a representation of $\pi_1(T-\Delta, t_o)$; this kind of representation goes under the general heading of "monodromy" .

We shall describe now some examples to clarify this discussion.

a) The family of hypersufaces of degree d in $\mathbb{P}_n$.

Let A denote the vector space of homogeneous polynomials of degree d in n+1 variables (over $\mathbb{R}$ or $\mathbb{C}$). To each $f \in A - \{0\}$ one can associate its locus of zeros $F(f)$ (a hypersurface of degree d) in P^n ; the family of these hypersurfaces can be seen as a "continuous family" of subspace of $\mathbb{P}_n$ in the following way : first of all remark that $F(f)$ depends only on the "direction" of $f \in A$, so that the natural parameter space is not $A - \{0\}$ but the set of straight lines through the origin in A , i.e. the projective space $\mathbb{P}(A)$; denote by $\bar{f}$ the image of $f \in A - \{0\}$ in $\mathbb{P}(A)$. Define $X = \mathbb{P}_n$, $T = \mathbb{P}(A)$, $\tilde{X} = \left\{(x,\bar{f}) \in \mathbb{P}_n \times \mathbb{P}(A) \mid f(x) = 0\right\}$; the projections of the product space $\mathbb{P}_n \times \mathbb{P}(A)$ onto its factors induce by restriction to $\tilde{X}$' two maps $\tilde{X} \xrightarrow{\pi} T$, $\tilde{X} \xrightarrow{\varphi} X$. One can give these spaces the appropriate structure (algebraic or analytic or differentiable). In each case φ embeds every fibre $\pi^{-1}(t)$ into $X = \mathbb{P}_n$. Now we shall distinguish two cases:

a_1) $\mathbb{P}_n = \mathbb{P}_n(\mathbb{C})$. Let $\Sigma = \left\{x \in \tilde{X} \mid \pi \text{ does not have maximal rank at } x\right\} = \left\{x \in \tilde{X} \mid \text{the fibre of } \pi \text{ passing through } x \text{ has a singular point at } x\right\}$ and $\Delta = \pi(\Sigma) = \left\{t \in T \mid \pi^{-1}(t) \text{ has some singular point}\right\}$. One can show (by elimination theory) that Δ is an algebraic hypersurface in $\mathbb{P}(A)$, and in fact it is the locus of zeros of a homogeneous polynomial of degree $(n+1)\cdot(d-1)^n$.

Now Δ has real codimension two in T , so that $T - \Delta$ is connected . The fibres of π over $T - \Delta$ are exactly the non singular hypersurfaces of degree d in $\mathbb{P}_n(\mathbb{C})$; in general (for example $n = 2$, $d \geq 3$) there is no open set U in $T - \Delta$ such that the fibres $(\pi^{-1}(t))_{t\in U}$ are isomorphic (algebraically or complex analitically) to each other. On the other hand, if one considers the differentiable structure (or the real analytic one), since $\pi : \tilde{X} - \pi^{-1}(\Delta) \longrightarrow T - \Delta$ is a proper morphism between differentiable manifolds which has maximal rank everywhere, by a standard theorem in differential topology, every point $t_0 \in T - \Delta$ has an open neighbourhood U such that there exists a diffeomorphism of $\pi^{-1}(U)$ with $U \times \pi^{-1}(t_0)$ which commutes with the projections on U . This implies that π induces a fibre bundle over $T - \Delta$ with fibre $\pi^{-1}(t_0)$ and structure group the group of diffeomorphism of $\pi^{-1}(t_0)$ onto itself.

Now we consider the family of singular hypersurfaces, $(X_t)_{t\in\Delta}$. Let $\Gamma = \left\{\text{singular set of } \Delta\right\}$. Then one can show that $\Delta - \Gamma$ is connected and that each fibre X_t , $t \in \Delta - \Gamma$ has just one singular point, which is a generic quadratic point, i.e. given locally by $\sum_{1}^{n} x_i^2 = 0$. Again π induces a differentiable (or a real analytic) fibre bundle over $\Delta - \Gamma$ where the fibre is a "space with singularities" . In the next step, taking $\Gamma' = \{$singular set of $\Gamma\}$ one finds that $\Gamma - \Gamma'$ is no longer connected; nevertheless one has again a differentiable fibre bundle. In this way one stratifies, step by step, the family $(X_t)_{t\in T}$ but in general one cannot hope to get local differentiable trivializations as in the first step. For example let $n = 2$, $d = 4$. Then the family of four distinct lines through one point, contains a continuous set of non isomorphic elements from the differentiable point of view (in fact classified by the cross ratio); what one can ask for in general is only a topological trivialization. Now we return to the fibre bundle induced by π over $T-\Delta$. If one fixes an integer $r \geq 0$, one can find a representation $\sigma : \pi_1(T-\Delta, t_0) \longrightarrow \mathrm{Aut}\, H^r(X_{t_0}, \mathbb{Z})$; as we shall see this is interesting only for $r = n - 1$. But it is important to notice that this kind of monodromy associated with this continuous family does not depend on the map φ , i.e. it "forgets" the fact that the X_t are subspaces of X. Another kind of monodromy which takes care of this can be defined in the following way: let $\mathcal{E}$ be the space of all differentiable subspaces in X diffeomorphic to X_{t_0}; define a fundamental system of neighbourhoods of a point $E \in \mathcal{E}$ by fixing a tubulor neighbourhood U of E and a retraction (by geodesics for some Riemanian metrix on X) onto E: the neighbourhood of E is the set of all $E' \in \mathcal{E}$ contained in U and such that the restriction on E induces or diffeomorphism of E with E'. Denote by Ω the connected component of X_{t_0} in $\mathcal{E}$. Then one obtains a homomorphism $\bar{\sigma} : \pi_1(T - \Delta, t_0) \longrightarrow \pi_1(\Omega, X_{t_0})$. Obviously $\bar{\sigma}$ contains much more information than σ, but in general $\bar{\sigma}$ is very difficult to deal with. We shall return to the difference between σ and $\bar{\sigma}$ in example b).

a_2) $\mathbb{P}_n = \mathbb{P}_n(\mathbb{R})$. If one makes an algebraic computation by elimination theory one finds a discriminant polynomial that is the same as in example a_1), and whose locus of zeros is not just the set of critical values of the map $\pi : \tilde{X} \longrightarrow X$ but a larger one (in fact it also contains all the $\bar{f} \in \mathbb{P}(A)$ such that $\{x \in \mathbb{P}_n(\mathbb{C}) \mid f(x) = 0\}$ has singularities even though these may all have strictly complex coordinates" i.e. $\{x \in \mathbb{P}_n(\mathbb{R}) \mid f(x) = 0$ is non singular $)\}$.

Nevertheless $\Delta = \{$critical values of $\pi\}$ is a semianalytic set in T, which in general disconnects T. On each connected component of $T - \Delta$, one has a differentiable fibre bundle, the fibre in general being different (possibly empty) over different components. One can show that over the simple points of Δ, the fibre X_t has exactly one singular point, which is a generic quadratic point, i.e. of local equation $\sum_1^r x_i^2 = \sum_{r+1}^n x_i^2$ for some r.

For $n = 2$, $d = 2$, i.e. the case of real conics, one has $T \simeq \mathbb{P}_5$, Δ has degree three and $T - \Delta$ has two connected components Ω_1, Ω_2; over Ω_1 the fibre is empty, over Ω_2 the fibre is a circle. So $H_1(x_{t_0}, \mathbb{Z}) \simeq \mathbb{Z}$ and the continuous family obtained by "translating" the center of a circle along a stright line, changes the orientation of the fibre. Thus one gets a homomorphism $\pi_1(\Omega_2, t_0) \longrightarrow \mathrm{Aut}\, M_1(X_{t_0}, \mathbb{Z}) \simeq \mathbb{Z}/_{2\mathbb{Z}}$ which is in fact an isomorphism.

2. Germs of analytic spaces.

Let (X,x) be a gèrm of an analytic space, i.e. an analytic space X (with structure sheaf say $\mathcal{O}_X$) with a point $x \in X$, where one is interested only in the behaviour of X in an "arbitrary small neighbourhood" of x . One knows that the germ (X,x) is completely determined by the local ring $\mathcal{O}_{X,x}$; in fact the category of analytic spaces and analytic morphisms is equivalent (via the contravariant functor $(X,x) \rightsquigarrow \mathcal{O}_{X,x}$) to the category of $\mathbb{C}$ - analytic algebras (i.e. $\mathbb{C}$ - algebras which are isomorphic with some quotient $\mathbb{C}\{x_1,\ldots,x_n\}/_I$, where I is an ideal in the convergent power series ring in n variables $\mathbb{C}\{x_1,\ldots,x_n\}$) and local homomorphisms.

The Zariski tangent space to (X,x) is defined as the dual $T(X,x)$ of the vector space (over $\mathbb{C}$) $\mathfrak{m}/_{\mathfrak{m}^2}$, where $\mathfrak{m}$ denotes the maximal ideal of $\mathcal{O}_{X,x}$; its dimension coincides with the minimal integer n such that some neighbourhood of x in X can be embedded in $\mathbb{C}^n$, as one can see by the implicit function theorem. In fact let

(X,x) be embedded in $(\mathbb{C}^N,0)$ and denote by I its defining ideal in $\mathbb{C}\{x_1,\dots,x_N\}$; then the Zariski tangent space to (X,x) can be identified with the set $\{z \in \mathbb{C}^N \mid df(z) = 0 \text{ for all } f \in I\}$. If for some $f \in I$, df is not identically zero on $\mathbb{C}^N$, that means that $f = 0$ is a smooth manifold of dimension $N - 1$ that contains X. Here is another way to describe $T(X,x)$. Let T_1 be the space associated to the analytic algebra $\mathbb{C}\{t\}/(t^2)$. T_1 is a unreduced space consisting of one point and can be considered as the "first order neighbourhood" of 0 in $\mathbb{C}$ or as a "point with a direction".

Let x be a point in the analytic space X ; there is a canonical bijection between morphisms $\sigma : T_1 \to X$ s.th. $\sigma(0) = x$ and local homomorphism $\sigma^* : \mathcal{O}_{X,x} \to \mathcal{O}_{T_1} = \mathbb{C}\{t\}/(t^2)$. Moreover every morphism $\tau : \mathcal{O}_{X,x} \to \mathbb{C}\{t\}/(t^2)$ induces a $\mathbb{C}$ - linear map $\tau' : \mathfrak{m}_{X,x}/\mathfrak{m}^2_{X,x} \to \mathfrak{m}_{T_1}/\mathfrak{m}^2_{T_1} \simeq \mathbb{C}$, i.e. an element in $T(X,x)$; it is easy to see that $\tau \to \tau'$ is in fact a bijection so that $T(X,x)$ can be thought of as the set of all morphisms $\sigma : T_1 \to X$ s.th. $\sigma(0) = x$.

If $\varphi : (X,x) \to (Y,y)$ is an analytic morphism, to each $\sigma : T_1 \to (X,x)$ one associates $\varphi \circ \sigma : T_1 \to (Y,y)$; the linear map $d\varphi : T(X,x) \to T(Y,y)$ obtained in this way is called the differential of φ.

Let $\varphi : X \to T$ be a morphism between analytic spaces, $t \in T$. The set $X_t = \varphi^{-1}(t)$ has a natural structure of an analytic space, defined in the following way : for $x \in \varphi^{-1}(t)$ define I_x as the ideal in $\mathcal{O}_{X,x}$ generated by $\varphi_x^*(\mathfrak{m}_{T,t})$ where $\varphi_x^* : \mathcal{O}_{T,t} \to \mathcal{O}_{X,x}$ is the local homomorphism induced by φ ; for $x \in X - X_t$ define $I_x = \mathcal{O}_{X,x}$. The collection of all the I_x forms a coherent ideal sheaf in $\mathcal{O}_X$ defining on X_t the structure of an analytic space, which will be called the fibre of φ over t. In general, even if X and T are reduced, the fibre X_t may not be reduced.

Remark that if $\varphi : (X,x) \to (T,t)$ is a morphism between germs, then only the fibre X_t over t can be defined and it is a germ of a space. One has

$$\mathcal{O}_{X_t,x} = \mathcal{O}_{X,x} / \varphi^*(\mathfrak{m}_{T,t}) \cdot \mathcal{O}_{X,x} .$$

3. Deformations.

Let $\varphi : X \rightarrow T$ be a morphism between analytic spaces, $x \in X$, $t = \varphi(x)$. Then $\varphi_x^* : \mathcal{O}_{T,t} \rightarrow \mathcal{O}_{X,x}$ makes $\mathcal{O}_{X,x}$ into an $\mathcal{O}_{T,t}$ module; φ is said to be flat at x if $\mathcal{O}_{X,x}$ is a flat $\mathcal{O}_{T,t}$ module (see the definition in [3], I.2.3.) .

The set of points where φ is flat is open in X and in fact it coincides with the complement of an analytic subset in X ([6], IV. 9). The geometric meaning of flatness is explained by the following theorem.

Theorem ([11], Satz 1). A flat morphism is an open map. Vice versa let $\varphi : X \rightarrow T$ be an analytic morphism with reduced fibres between reduced complex spaces, T non singular

i) if φ is an open map, φ is flat

ii) if for all $x \in X$ there exist an open set U containing x s.th. $\dim_y \varphi^{-1}\varphi(y)$ does not depend on $y \in U$, φ is flat.

So, in some sense, a flat map $\varphi : X \to T$ can be considered as a family of analytic spaces $(X_t)_{t \in T}$ of constant dimension.

Remark that if T is singular, the condition " the dimension of the fibre is locally constant" does not imply that φ is flat. For example let $T = \{(x,y) \in \mathbb{C}^2 | x^2=y^3\}$, $\varphi: \mathbb{C} \to X$ defined by $\varphi(t) = (t^3, t^2)$; then φ is a homeomorphism (the normalization of T) but it is not flat at 0 . This "suggests" that a flat map is a "good analytic family of analytic spaces" .

In the global theory one usually assumes that φ is also proper; here we are interested in the local theory so that we shall work with germs of analytic spaces.

<u>Definitions</u>. A <u>singularity</u> is a germ of an analytic space. A <u>family of singularities</u> is a flat morphism $\varphi: (X,x) \to (T,t)$. Let (X_0,x) be a singularity; a <u>deformation</u> of (X_0,x) is a family of singularities $\varphi: (X,x) \to (T,t)$ together with an identification j of (X_0,x) with the fibre $(X_{t,x})$ of φ ; φ is called a deformation of (X_0,x) over (T,t) .

Let $\varphi : (X,x) \rightarrow (S,s)$ be a family of singularities and $\sigma : (T,t) \rightarrow (S,s)$ be any morphism. Taking the fibre product of φ and σ one finds a flat morphism $\varphi' : (X',x') \rightarrow (T,t)$ which is called the pull back of the family φ by the basis extension σ. Moreover, if (X_0,x) is identified with the fibre of φ, there is a canonical identification of (X_0,x) with the fibre of φ'.

Hence if $\varphi : (X,x) \rightarrow (S,s)$ is a deformation of (X_0,x) over (S,s), any map $\sigma : (T,t) \rightarrow (S,s)$ induces a deformation of (X_0,x) over (T,t).

From now on we shall aften denote a germ (X,x) simply by X, if no confusion is possible.

Let $Def(X_0,T)$ denote the set of all (isomorphism classes of) deformations of the singularity X_0 over T. We have seen that $Def(X_0,-)$ can be considered as a controvariant functor from the category of germs of analytic spaces to the category of sets.

One can ask if this functor is representable, i.e. if there exists a deformation $\varphi : X \rightarrow S$ of X_0 (<u>the universal deformation</u> of X_0) such that for all germs of spaces S', the pull back of φ induces a bijection

$Morf(S',S) \to Def(X_0,S')$; in other words if any other deformation $\varphi' : X' \to S'$ of X_0 is induced by a unique $\sigma : T \to S$.

However, as we shall see later, the universal deformation does not exist in general, even for simple singularities.

So one ask for something less then universality: a deformation $\varphi : X \to S$ of X_0 is called a <u>versal</u> deformation of X_0 if any other deformation $\varphi' : X' \to S'$ of X_0 is induced by some morphism $\sigma : S' \to S$, i.e. one does not require σ to be unique ; φ will be called <u>semiuniversal</u> if it is versal and if the differential $d\sigma : T(S',s') \to T(S,s)$ is uniquely determined by φ' . Obviously a versal deformation can be too big : if $\varphi: X \to S$ is versal for X_0 and Y is any space, $\varphi \times id_Y : X \times Y \to S \times Y$ will also be versal for X_0 . On the other hand :

<u>Proposition</u> 1. If the semiuniversal deformation of a singularity exists, it is uniquely determined up to an isomorphism .

Proof. Let $\varphi : X \to S$, $\varphi' : X' \to S'$ be semiuniversal deformations of X_0 . There exist $\sigma : S \to S'$ inducing φ and $\tau : S' \to S$ inducing φ' , so that $\tau \circ \sigma : S \to S$ induces φ .

Since φ is semiuniversal, it follows that $d(\tau \circ \sigma) =$ $=$ d(identity map on S) = identity on $T(S,s)$; this implies that $\tau \circ \sigma$ is an isomorphism. The same for $\sigma \circ \tau$, so that σ and τ are isomorphisms.

As we have seen the Zariski tangent space $T(S,s)$ of a space S at a point s is identified with the set of morphisms $T_1 \to (S,s)$. If $\varphi : X \to S$ is a deformation of X_0 , each element in $T(S,s)$ induces a deformation of X_0 over T_1 ; in this way one obtains a map $T(S,s) \to$ $\to \mathrm{Def}(X_0,T_1)$.

Proposition 2. i) Let $\varphi : X \to S$ be the semiuniversal deformation of X_0 . Then the associated map $T(S,s) \to$ $\to \mathrm{Def}(X_0,T_1)$ is bijective

ii) Let $\varphi : X \to S$ be a deformation of X_0 s.th. S is smooth and the associated map $T(S,s) \to \mathrm{Def}(X_0,T_1)$

is bijective.

If X_0 admits a semiuniversal deformation, then is semiuniversal.

Proof. i) If two morphisms σ, $\sigma' : T_1 \to (S,s)$ induce the same element in $Def(X_0,T_1)$, then by semiuniversality of φ one gets $d\sigma = d\sigma'$; this implies $\sigma = \sigma'$ since the domain is T_1, so that $T(S,s) \to$ $\to Def(X_0,T_1)$ is injective.

Now let $\tau : X' \to T_1$ be any element in $Def(X_0,T_1)$; by the semiuniversality of φ, τ is induced by a morphism $T_1 \to (S,s)$ and hence by an element of $T(S,s)$.

ii) Let $\varphi' : X' \to S'$ be a semiuniversal deformation of X_0, and let $\sigma : S \to S'$ induce φ. The hypothesis that $T(S,s) \to Def(X_0,T_1)$ is bijective implies that $d\sigma : T(S,s) \to T(S',s')$ is an isomorphism; since S is smooth this implies that σ is biholomorphic.

With the same arguments one proves the following :

Proposition 3. i) Let $\varphi : X \to S$ be a versal deformation of X_o. Then the associated map $T(S,s) \to \to Def(X_o, T_1)$ is surjective.

ii) Let $\varphi : X \to S$ be the semiuniversal deformation of X_o and $\varphi' : X' \to S'$ a deformation of X_o s.th. S' is smooth. Then S is smooth and $\varphi' = \varphi \times id_{\mathbb{C}^h} : X \times \mathbb{C}^h \to S \times \mathbb{C}^h$ for some integer $h \geq 0$.

4. <u>Infinitesimal deformations</u>.

We shall compute $\mathrm{Def}(X_0, T_1)$ in terms of $\mathcal{O}_{X_0}$. In particular we shall see that it has a natural structure of a vector space over $\mathbb{C}$: if $(X,x) \to (S,s)$ is a deformation of X_0, the associated map $T(S,s) \to \mathrm{Def}(X_0,T_1)$ will be a linear map. This will imply that a singularity X_0 s.th. $\mathrm{Def}(X_0,T_1)$ is not a finite dimensional vector space, does not admit semiuniversal deformations. On the other hand this will indicate to us whether a given deformation is semiuniversal or not.

First of all we look at the local analytic form of the deformation of a singularity embedded in $\mathbb{C}^n$

We recall the following two lemmas:

<u>Lemma 1</u>. ([7], satz 2.2) Let $\varphi : (X,x) \to (T,t)$ be a morphism between germs of analytic spaces, $X_t = \varphi^{-1}(t)$ the fibre of φ over t. If X_t is embedded in $\mathbb{C}^n$, there exists an embedding of X in $\mathbb{C}^n \times T$ s.th. φ is induced by the canonical projection $\mathbb{C}^n \times T \to T$.

Lemma 2. ([16], (1.7)) Let $X \xrightarrow{\sigma} X'$, $\varphi: X \to T$, $\varphi': X' \to T$ be a commutative diagramm of morphisms between germs of analytic spaces.

If φ is flat and σ induces an isomorphism between the fibres $\varphi^{-1}(t)$ and $\varphi'^{-1}(t)$, then σ is an isomorphism.

We can now give the following analytic description of the deformations of X_0 :

Lemma 3. Let X_0 be embedded in $\mathbb{C}^n$ and let $I_0 \subset \mathbb{C}\{x_1,\ldots,x_n\}$ be its defining ideal, generated say by $f_1,\ldots,f_m \in \mathbb{C}\{x_1,\ldots,x_n\}$. If $\varphi : X \to T$ is a deformation of X_0, one may suppose that :

i) X is defined in $\mathbb{C}^n \times T$ by an ideal $I \subset \mathcal{O}_{\mathbb{C}^n \times T} = \mathcal{O}_T\{x_1,\ldots,x_n\}$, $I_0 = I \cap (t_1,\ldots,t_r)$ (where $t_1,\ldots,t_r$ are generators of the maximal ideal of $\mathcal{O}_T$) and the morphism φ is the restriction to X of the canonical projection $\mathbb{C}^n \times T \to T$.

ii) If $F_1,\dots,F_m \in I$ are such that $F_i|_{\mathbb{C}^n} = f_i$, $i = 1,\dots,m$, then I is generated by $F_1,\dots,F_m$.

So any deformation of X_0 can be found in the following way : Take $F_1,\dots,F_m \in \mathcal{O}_T\{x_1,\dots,x_n\}$ of the form $F_i(x,t) = f_i(x) + \sum_1^r G_{ij}(x,t)\cdot t_j$, $i = 1,\dots,m$, and let X be defined in $\mathbb{C}^n \times T$ by the ideal generated by the F_i. Require that the restriction φ to X of the canonical projection $\mathbb{C}^n \times T \to T$ is flat.

Remark that the flatness of φ is equivalent to the following fact :
"every relation $\sum_1^m p_i f_i = 0$ in $\mathbb{C}\{x_1,\dots,x_n\}$ can be lifted to a relation $\sum_1^m P_i F_i = 0$ in $\mathcal{O}_T\{x_1,\dots,x_n\}$ (lifted means that $P(x,t) = p_i(x) + \sum_1^r H_{ij}(x,t)\cdot t_j$) " .

The structure of $Def(X_0,T_1)$.

Remark first that the $\mathcal{O}_{\mathbb{C}^n}$ module I_0/I_0^2 can be considered as a $\mathcal{O}_{X_0}$ module; moreover it does not depends on the particular embedding $X_0 \hookrightarrow \mathbb{C}^n$ but only on X and n .

We shall now construct a bijection between the dual of the $\mathcal{O}_{X_o}$ - module I_o/I_o^2 and the deformations $X \to T_1$ of X_o where X is embedded in $\mathbb{C}^n \times I_1$ and φ is induced by the projection $\mathbb{C}^n \times T_1 \to T_1$.

Let $\sigma : I_o/I_o^2 \to \mathcal{O}_{X_o}$ be a homomorphism of $\mathcal{O}_{X_o}$ modules: choose $g_1,\ldots,g_m \in \mathcal{O}_{\mathbb{C}^n}$ such that the image of g_i in $\mathcal{O}_{X_o}$ coincide with σ (image of f_i in I_o/I_o^2) , $i = 1,\ldots,m$.

Define $X \hookrightarrow \mathbb{C}^n \times T_1$ by the ideal $I = (F_1,\ldots,F_m)$ where $F_i = f_i + t\, g_i$, $i = 1,\ldots,m$. We claim that the projection map $\mathbb{C}^n \times T_1 \to T_1$ induces a flat morphism $\varphi\colon X \to T_1$. As we have seen we need only check that every relation $\sum_1^m p_i f_i = 0$ can be lifted to a relation $\sum_1^m P_i F_i = 0$. In fact let $\sum_1^m p_i f_i = 0$. Then, since σ is a homomorphism, $\sum_1^m p_i g_i$ vanishes on X_o , so that $\sum_1^m p_i g_i = \sum_1^m q_i f_i$; define $P_i = p_i - t\, q_i$. Then

$$\sum_1^m P_i F_i = \sum_1^m (p_i - t\, q_i)\cdot(f_i+t\, g_i) = \sum_1^m p_i f_i +$$

$$+ t\left(\sum_1^m p_i g_i - \sum_1^m q_i f_i\right) = 0 \text{ , so that } \varphi \text{ is flat.}$$

Moreover one can verify that X (that is the ideal $(F_1,\ldots,F_m)$) depends only on σ .

On the other hand, let $\varphi : X \to T_1$ be a deformation of X_o , s.th. X is embedded in $\mathbb{C}^n \times T_1$ and φ is induced by the projection $\mathbb{C}^n \times T_1 \to T_1$. Then X is defined by an ideal $I = (F_1,\ldots,F_m)$ in $\mathcal{O}_{T_1}\{x_1,\ldots,x_n\}$ where $F_i = f_i + t\, g_i$, $g_i \in \mathbb{C}\{x_1,\ldots,x_n\}$, $i = 1,\ldots,m$.

Let $\bar{g}_i$ denote the image of g_i in $\mathcal{O}_{X_o}$, and define

$$\sigma : I_o/I_o^2 \to \mathcal{O}_{X_o} \quad \text{by} \quad f_i \rightsquigarrow \bar{g}_i .$$

We must verify that σ is well defined. In fact let $\sum_1^m p_i f_i = 0$ in I_o ; then by flatness of φ one gets

$\sum_1^m (p_i+tq_i)\cdot(f_i+tg_i) = 0$, i.e. $\sum_1^m (q_i f_i + p_i g_i) = 0$, and hence $\sum_1^m p_i \bar{g}_i$ is zero in $\mathcal{O}_{X_o}$.

Moreover $\sigma(f_i f_j) = f_i \cdot \sigma(f_j) = f_i \cdot \bar{g}_j$ is zero in $\mathcal{O}_{X_o}$ so that σ vanishes on I_o^2 . Also one can verify that σ depends only on X .

Now we want to know when two homomorphism $I_o/I_o^2 \to \mathcal{O}_{X_o}$ induce the same deformation.

Denote by $\Omega^1_{\mathbb{C}^n}$ the $\mathcal{O}_{\mathbb{C}^n}$ module of germs of holomorphic 1-forms at $0 \in \mathbb{C}^n$. Then $\Omega^1_{\mathbb{C}^n} / I_0 \cdot \Omega^1_{\mathbb{C}^n}$ can be considered as an $\mathcal{O}_{X_0}$ module.

Moreover differentiation induces an homomorphisms of $\mathcal{O}_{X_0}$ modules $d : I_0/I_0^2 \to \Omega^1_{\mathbb{C}^n} / I_0 \Omega^1_{\mathbb{C}^n}$.

<u>Proposition</u> 4. The transpose d^* of d induces an exact sequence

$$\mathrm{Hom}_{\mathcal{O}_{X_0}}\left(\Omega^1_{\mathbb{C}^n} / I_0 \Omega^1_{\mathbb{C}^n}, \mathcal{O}_{X_0}\right) \xrightarrow{d^*} \mathrm{Hom}_{\mathcal{O}_{X_0}}\left(I_0/I_0^2, \mathcal{O}_{X_0}\right) \to \mathrm{Def}(X_0, T_1) \to 0$$

In other words $\mathrm{Def}(X_0, T_1)$ can be identified with coker d^* .

<u>Proof</u>. Suppose that $\varphi, \varphi' : I_0/I_0^2 \to \mathcal{O}_{X_0}$ induce isomorphic deformations. Let φ be represented by $g_i \in \mathbb{C}\{x_1, \ldots, x_n\}$ and φ' by $g_i' \in \mathbb{C}\{x_1, \ldots, x_n\}$, $i = 1, \ldots, m$. Then there exists an isomorphism of $\mathbb{C}^n \times T$ onto itself that transforms X in X' . This isomorphism must be of the form $(z,t) \to (w(z,t),t) = (z+t\, w(z),t)$,

because it must respect the projections onto T .

So for all i : $f_i + t\, g_i' \in (f_i(z+t\cdot w(z)) +$ $+ t\, g_i(z+tw(z)) = (f_i(z) + t(g_i(z) + df_i\cdot w(z)))$. For $h = 1,\dots,m$ one gets

$$f_h + t\, g_h' = \sum_1^m (a_i+t\, b_i)\cdot(f_i+t(g_i+df_i\cdot w)) =$$

$$= \sum_1^m a_i f_i + t\left(\sum_1^m a_i\, g_i + b_i\, f_i + \left(\sum_1^m a_i\, df_i\right)\cdot w\right) .$$

By flatness $f_h = \sum_1^m a_i\, f_i$ implies $\sum_1^m a_i g_i =$ $= g_h + \sum_1^m a_i'\, f_i$ so that

$$g_h' = g_h + f + \sum_1^m (a_i df_i)\cdot w \quad , \quad f \in I_0 \quad .$$

Moreover $f_h = \sum_1^m a_i\, f_i$ implies $df_h = \sum_1^m a_i\, df_i +$ $+ f'$, $f' \in I_0$ so that

$$g_h' = g_h + f'' + df_h\cdot w \quad , \quad f'' \in I_0 \quad .$$

Since g_h' was chosen up to additive terms in I_0 one may suppose $g_h' = g_h + df_h\cdot w$. So the difference between φ and φ' is of the form $f_h \rightsquigarrow df_h \rightsquigarrow df_h\cdot w$.

It is a matter of simple computations to finish the proof.

Remark. Let X_0 be a submanifold of a manifold M, $\mathcal{J} = \mathcal{J}_{X_0}$ the ideal sheaf in $\mathcal{O}_M$ defining X_0. Then $\mathcal{J}/\mathcal{J}^2$ can be considered as a coherent $\mathcal{O}_{X_0}$-module and in fact free of rank $\dim M - \dim X_0$. Its dual over $\mathcal{O}_{X_0}$ is isomorphic (perhaps by definition) with the normal bundle of $X_0 \to X$, so that the elements of the sheaf $\mathrm{Hom}_{\mathcal{O}_{X_0}}(\mathcal{J}/\mathcal{J}^2, \mathcal{O}_{X_0})$ can be interpreted as germs of normal vector fields along X_0.

In the same way the dual of $\Omega^1(M)/\mathcal{J}\Omega^1(M)$ over $\mathcal{O}_{X_0}$ can be interpreted as the germs of vector fields along X_0 tangent to M. This shows that

$$d_x^* : \mathrm{Hom}_{\mathcal{O}_{X_0}}\left(\Omega^1(M)/\mathcal{J}\Omega^1(M), \mathcal{O}_{X_0}\right)_x \to \mathrm{Hom}_{\mathcal{O}_{X_0}}\left(\mathcal{J}/\mathcal{J}^2, \mathcal{O}_{X_0}\right)_x$$

is surjective if X_0 is non singular at x, since every normal vector field along X_0 can be interpreted as a vector field tangent to M.

On the other hand if X_0 is singular at x, it is possible that the elements of $\mathrm{Hom}_{\mathcal{O}_{X_0}}\left(\mathcal{J}/\mathcal{J}^2, \mathcal{O}_{X_0}\right)$ cannot be interpreted as vector fields along X_0, so d_x^* may not be surjective. As an example consider $X_0 \hookrightarrow \mathbb{C}^2$ defined by $f(x,y) = x\cdot y = 0$. Then $\mathrm{Hom}_{\mathcal{O}_{X_0}}\left(\mathcal{J}/\mathcal{J}^2, \mathcal{O}_{X_0}\right) \simeq \mathcal{O}_{X_0}$;

with this identification.

The function $1 \in \mathcal{O}_{X_0}$ defines at each regular point of X_0 a normal vector, but one cannot extend this field to $0 \in X_0$ (in fact any of these vector fields has limit ∞ at 0).

Since coker d^* is a coherent $\mathcal{O}_X$ module and is concentrated at the singular points of X_0 one gets:

Corollary. Let (X_0, x) be a germ of an analytic space with an isolated singularity at X_0 . Then $\mathrm{Def}(X_0, T_1)$ is a finite dimensional vector space.

Proposition 5. Let $(X_0, 0)$ be a hypersurface in $(\mathbb{C}^n, 0)$ defined by $f \in \mathbb{C}\{x_1, \ldots, x_n\}$. Then $\mathrm{Def}(X_0, T_1)$ can be identified with $\mathbb{C}\{x_1, \ldots, x_n\} \Big/ \left(f, \frac{\partial f}{\partial x_1}, \ldots, \frac{\partial f}{\partial x_n}\right)$.

In particular if X_0 has a non isolated singularity at 0 , it does not have semiuniversal deformations.

<u>Proof.</u> I_o/I_o^2 can be identified with $\mathcal{O}_{X_o} = \mathbb{C}\{X_1,\dots,X_n\}/(f)$, and $\Omega^1_{\mathbb{C}^n}/I_o\Omega^1_{\mathbb{C}^n}$ with $\mathcal{O}^n_{X_o} = \left(\mathbb{C}\{x_1,\dots,x_n\}/(f)\right)^n$. Moreover $d : I_o/I_o^2 \longrightarrow \Omega^1_{\mathbb{C}^n}/I_o\Omega^1_{\mathbb{C}^n}$ considered as a map $\mathcal{O}_{X_o} \to \mathcal{O}^n_{X_o}$ is defined by $1 \to \left(\frac{\partial f}{\partial x_1},\dots,\frac{\partial f}{\partial x_n}\right)$ so that its transpose is the map $\mathcal{O}^n_{X_o} \to \mathcal{O}_{X_o}$ defined by $(\alpha_1,\dots,\alpha_n) \to \sum_1^n \alpha_i \frac{\partial f}{\partial x_i}$, whose cokernel is

$$\mathbb{C}\{x_1,\dots,x_n\}/(f)\Big/\left(\frac{\partial f}{\partial x_1},\dots,\frac{\partial f}{\partial x_n}\right) =$$

$$= \mathbb{C}\{x_1,\dots,x_n\}\Big/\left(f,\frac{\partial f}{\partial x_1},\dots,\frac{\partial f}{\partial x_n}\right).$$

Now we want to explicitate the homomorphism $\bar{\varphi} : T(S,s) \to Def(X_o,T_1)$ associated to a deformation $\varphi : X \to S$ of an isolated singularity of hypersurface type $X_o \hookrightarrow \mathbb{C}^n$.

As before we may suppose that X is defined in $\mathbb{C}^n \times S$ by a function $F(x,s) \in \mathcal{O}_S\{x_1,\dots,x_n\}$, such that

$F(x,0) = f(x)$ defines X_0 in $\mathbb{C}^n$.

Let $\mathfrak{m}$ be the maximal ideal in $\mathcal{O}_S$ and denote by $T(S)$ (the tangent space of S) the subspace of S defined by $\mathfrak{m}^2$, i.e. $\mathcal{O}_{T(S)} = \mathcal{O}_{S/\mathfrak{m}^2}$. The embedding $T(S) \to S$ (associated to $\mathcal{O}_S \to \mathcal{O}_{S/\mathfrak{m}^2}$) induces a deformation $\tilde{\varphi} : \tilde{X} \to T(S)$. The image $\tilde{F}$ of F in $\mathcal{O}_{T(S)}\{x_1,\ldots,x_n\}$ is constructed by differentiating F in s (i.e. expanding F in power serie in s); in other words one has $\tilde{F}(x,s) = F(x,0) + s\cdot G(x,0)$ where $s.G(x,0) \in \mathfrak{m}/\mathfrak{m}^2\{x_1,\ldots,x_n\}$, so that for all linear forms $\nu : \mathfrak{m}/\mathfrak{m}^2 \to \mathbb{C}$ (i.e. ν a tangent vector to S) one gets an element $\frac{\partial}{\partial \nu} F(x,0) \in \mathbb{C}\{x_1,\ldots,x_n\}$. The element in $\mathrm{Def}(X_0,T_1)$ associated to ν is defined by $F(x,0) + \frac{\partial}{\partial} F(x,0)\cdot t$, which induce the homomorphism $(f)/(f^2) \to \mathcal{O}_{X_0}$ defined by $f(x) \rightsquigarrow \frac{\partial}{\partial} F(x,0)|_{X_0}$. Hence $\bar{\varphi}$ can be written as $\nu \to \frac{\partial F}{\partial \nu}(x,0) \in \mathbb{C}\{x_1,\ldots,x_n\}/(f,\frac{\partial f}{\partial x_i})$.

5. Hypersurface singularities.

Let $(X_0,0)$ be a hypersurface in $(\mathbb{C}^n,0)$ defined by $f \in \mathbb{C}\{x_1,\ldots,x_n\}$; X_0 has an isolated singularity at 0 iff $\mathbb{C}\{x_1,\ldots,x_n\}/\left(\frac{\partial f}{\partial x_1},\ldots,\frac{\partial f}{\partial x_n}\right)$ is finite dimensional. This is equivalent to say that $\mathbb{C}\{x_1,\ldots,x_n\}/(f,\frac{\partial f}{\partial x_1},\ldots,\frac{\partial f}{\partial x_n})$ is finite dimensional and hence by proposition 5 to say that $\mathrm{Def}(X_0,T_1)$ is a finite dimensional vector space.

Let $g_1,\ldots,g_\tau \in \mathbb{C}\{x_1,\ldots,x_n\}$ be such that their images in $\mathbb{C}\{x_1,\ldots,x_n\}/\left(f,\frac{\partial f}{\partial x_1},\ldots,\frac{\partial f}{\partial x_n}\right)$ form a basis of that vector space over $\mathbb{C}$.

Let $X \subset \mathbb{C}^n \times \mathbb{C}^\tau$ be defined by $F(x,\alpha) = f(x) + \sum_1^\tau \alpha_i\, g_i(x) \in \mathbb{C}\{x_1,\ldots,x_n, \alpha_1,\ldots,\alpha_\tau\}$ and let $\varphi : X \to \mathbb{C}^\tau$ be the morphism induced by the projection $\mathbb{C}^n \times \mathbb{C}^\tau \to \mathbb{C}^\tau$. Then φ is flat and its fibre over the origin is canonically identified with X_0. Moreover the associated map $T(\mathbb{C}^\tau,0) \to \mathrm{Def}(X_0,T_1)$ is an isomorphism so that if X_0 has some semiuniversal deformation, then by proposition 2 , φ is semiuniversal.

Before giving the proof that φ is indeed semiuniversal, we shall show why we excluded universal deformations.

Example: let $(X_o,0)$ be defined in $(\mathbb{C},0)$ by $x^2 = 0$, i.e. $X_o \cong T_1$. Then X_o has no universal deformations. In fact as we have seen, if X_o has a semiuniversal deformation this must coincide with $\varphi: (\mathbb{C},0) \to (\mathbb{C},0)$ defined by $\varphi(x) = x^2$; so if a universal deformation exists it coincides with φ . But φ is not universal: in fact there exist commutative diagrams

$$\begin{array}{ccc} \mathbb{C} & \xrightarrow{\lambda} & \mathbb{C} \\ \varphi\downarrow & & \downarrow\varphi \\ \mathbb{C} & \xrightarrow{\sigma} & \mathbb{C} \end{array}$$

such that λ induces the identity on the fibre of φ over 0 , but σ is not the identity. To find such a diagram it is sufficient to start with any $\sigma(t) = t + a_2t^2 + \ldots + a_nt^n + \ldots$ in $\mathbb{C}\{t\}$ and to solve the equation $\lambda(t)^2 = \sigma(t^2)$ in $\lambda = t + \alpha_2t^2 + \ldots$ $\ldots + \alpha_nt^n + \ldots \in \mathbb{C}\{t\}$.

<u>Theorem 1</u> ([17]) Any isolated singularity of a hypersurface has a semiuniversal deformation.

<u>Proof.</u> As we have seen it is sufficient to show that the deformation of $X_o = \{f = 0\}$ defined by $F(x,\alpha) = f(x) + \sum_1^{\tau} \alpha_i g_i(x) = 0$, where $g_1, \ldots, g_\tau$ induce a basis of $\mathbb{C}\{x_1,\ldots,x_n\} / \left(f, \frac{\partial f}{\partial x_1}, \ldots, \frac{\partial f}{\partial x_n}\right)$, is a versal deformation. We shall only sketch the proof given in [10], where the case of an isolated singularity of a complete intersection is considered.

One has to show that if $\varphi' : X' \to S$ is any deformation of X_o, there exists $\sigma : S \to \mathbb{C}$ that induces φ'.

This problem can be formulated analytically in the following way :

One may suppose that $X' \hookrightarrow \mathbb{C}^n \times S$ is defined by a $G(x,s) \in \mathcal{O}_S\{x_1,\ldots,x_n\}$ s.th. $G(x,0) = f(x)$ and that φ' is induced by the projection $\mathbb{C}^n \times S \to S$.

Let S be embedded in $\mathbb{C}^r$. Then one wants to find functions :

i) $w(z,s) : \mathbb{C}^n \times S \to \mathbb{C}^n$

ii) $\alpha(s) : \mathbb{C}^r \to \mathbb{C}^s$

iii) $h(z,s) : \mathbb{C}^n \times \mathbb{C}^r — \mathbb{C}$

satisfying $h(z,0) = 0$ and

$$(*)_\infty \qquad G(w(z,s),s) = (1+ h(z,s))\cdot F(z, (s)).$$

This problem consists in finding convergent power series. The first step in the proof is to solve the problem formally, i.e. by formal power series.

This is not very hard: we consider w, α, h as power series in s, and we define w^μ, α^μ, h^μ to be their part of degree $\leq \mu$.

Then $(*)_\infty$ is equivalent to the sequence of conditions

$$(*)_\mu \qquad G(w^\mu(z,s),s) \equiv (1+ h^\mu(z,s))\cdot F(z,\alpha^\mu(s)) \quad \text{modulo } (s)^{\mu+1} .$$

Then one verifies that if $(*)_\mu$ has been solved, one can find homogeneous parts of degree $\mu+1$ of w,α,h so that to solve $(*)_{\mu+1}$.

In this way one constructs a formal solution.

Now it can happen that this is not a holomorphic

solution, i.e. that the power series we have found do not converge. The second step in the proof is to show that the recursive solutions to $(*)_\mu$ can be found with prescribed upper bound on the homogeneous terms on (good) fixed closed neighbourhoods of 0 in $\mathbb{C}^n \times \mathbb{C}^r$ and $\mathbb{C}^r$, so that the formal solutions converg.

<u>Remark</u>. a) When we choose $g_1, \ldots, g_\tau \in \mathbb{C}\{x\}$ in such a way that they induce a basis in $\mathbb{C}\{x\}/(f, \frac{\partial f}{\partial x_i})$, we may suppose $g_\tau = -1$ (if X_0 is really singular at x; if not $\tau = 0$). So $X \subset \mathbb{C}^n \times \mathbb{C}^\tau$ is defined by $f(x) + \sum_1^{\tau-1} \alpha_i g_i(x) = \alpha_\tau$. This shows that $(x_1, \ldots, x_n, \alpha_1, \ldots, \alpha_{\tau-1})$ can be used as local coordinates on X; it follows that the semiuniversal deformation of X_0 has the analytic form $(x_1, \ldots, x_n, \alpha_1, \ldots, \alpha_{\tau-1}) \longrightarrow$ $\longrightarrow (\alpha_1, \ldots, \alpha_{\tau-1}, f(x) + \sum_1^{\tau-1} \alpha_i g_i(x))$ where $g_1, \ldots, g_{\tau-1}$ induce a basis of $\mathfrak{m}/(f, \frac{\partial f}{\partial x_i})$, $\mathfrak{m}$ the maximal ideal in $\mathbb{C}\{x\}$.

b) the existence theorem 1 was proved first by

Tjurina [17]. After that Schlessinger [15] stated the existence of the formal semiuniversal deformation for any isolated singularity of an analytic space.

Finally Grauert [8] has given the general existence theorem for the (holomorphic) semiuniversal deformation of an isolated singularity.

6. <u>The geometry of the discriminant</u>.

Let $\varphi : (X,x) \to (S,s)$ be a deformation of the hypersurface $(X_0,0) \hookrightarrow (\mathbb{C}^{n+1},0)$. Then X can be embedded in $\mathbb{C}^{n+1} \times S$ so that φ is induced by the canonical projection $\mathbb{C}^{n+1} \times S \to S$. Let $f \in \mathbb{C}\{x_0,\ldots,x_n\}$ be a local equation for X_0 in $\mathbb{C}^{n+1}$; then X can be defined by a local equation $F(x,s) \in \mathcal{O}_{\mathbb{C}^{n+1}\times S} = \mathcal{O}_S\{x_0,\ldots,x_n\}$ such that $F|_{\mathbb{C}^{n+1}} = f$; in other words the homomorphism $\mathcal{O}_S\{x_0,\ldots,x_n\} \to \mathcal{O}_{S}/\mu_S\{x_0,\ldots,x_n\} = \mathbb{C}\{x_0,\ldots,x_n\}$ sends F to f.

The <u>critical space</u> $C \hookrightarrow X$ is defined as the subspace of X defined by the ideal $\mathcal{C} = \left(\frac{\partial F}{\partial x_0},\ldots,\frac{\partial F}{\partial x_m}\right) \subset$ $\subset \mathcal{O}_S\{x_0,\ldots,x_n\}$.

One can verify that C does not depend on the choice of F but only on φ .

Suppose X_0 has an isolated singularity at x . Then the fibre of $\varphi|_C : C \to S$ over $s \in S$ is just one (in general not reduced) point, so that by the Weierstrass preparation theorem, $\varphi|_C$ is a finite morphism. In particular $\varphi|_C$ is a locally proper morphism, so that the direct image sheaf $\mathcal{F}$ of $\mathcal{O}_C$ by $\varphi|_C$ is a coherent sheaf on S .

Let $\mathcal{O}_S^q \xrightarrow{\sigma} \mathcal{O}_S^p \longrightarrow \mathcal{F} \longrightarrow 0$ be a finite presentation of $\mathcal{F}$ and denote by $\mathcal{D}$ the ideal sheaf in $\mathcal{O}_S$ generated by the $(p \times p)$ minors of the matrix defining σ .

Again one can verify that $\mathcal{D}$ depends only on φ and not on the presentation σ ; the subspace D defined by $\mathcal{D}$ in S is called the <u>discriminant</u> of φ .

The discriminant is functorial in the sense that if $\varphi' : X' \to S'$ is a deformation of X_0 induced by a morphism $\lambda : S' \to S$, then the discriminant of φ' is just the analytic inverse image of D trough λ .

We shall see that if φ is the semiuniversal deformation of X_0 , $\mathcal{D}$ is a principal ideal sheaf; it will follow that the discriminant of any deformation of X_0 is defined by one local equation.

The sets underlying C and D are called respectively the <u>critical set</u> and the <u>discriminant locus</u> and denoted by the same letters C , D . The geometric interpretation of them is

$C = \left\{ x \in X \;\middle|\; \text{the fibre} \quad \varphi^{-1}\varphi(x) \quad \text{is singular at} \quad x \right\}$

$D = \left\{ s \in S \;\middle|\; \text{the fibre} \quad \varphi^{-1}(s) \quad \text{is singular somewhere} \right\}$.

Before studying the properties of C, D, let us notice that we may suppose $f \in \mathfrak{m}^3$, where $\mathfrak{m}$ is the maximal ideal in $\mathbb{C}\{x_1,\ldots,x_n\}$. In fact by the Morse lemma, one can write in suitable coordinates $f = \tilde{f}(x_1,\ldots,x_m) + x_{m+1}^2 + \ldots + x_n^2$ with $\tilde{f} \in \mathfrak{m}^3$; it is easy to see that $\tilde{f}$ has an isolated singularity at $0 \in \mathbb{C}^m$ and that f, $\tilde{f}$ have semiuniversal deformations with the same parameter space $\mathbb{C}^\tau$, the same critical set C and the same map $\varphi : C \longrightarrow D$.

As we have seen, φ has the following analytic form:

$$(\alpha_1,\ldots,\alpha_{\tau-1}, x_1,\ldots,x_n) \qquad (\alpha_1,\ldots,\alpha_{\tau-1}, f(x) + \sum_1^{\tau-1} \alpha_i\, g_i(x))$$

where $1, g_1,\ldots,g_{\tau-1} \in \mathbb{C}\{x_1,\ldots,x_n\}$ induce a basis of

$$\mathbb{C}\{x_1,\ldots,x_n\} \Big/ \left(f, \frac{\partial f}{\partial x_1},\ldots,\frac{\partial f}{\partial x_n}\right) .$$

In particular $g_1,\ldots,g_{\tau-1}$ must generate the maximal ideal $\mathfrak{m}$ in $\mathbb{C}\{x_1,\ldots,x_n\}$. The Jacobian of φ is

$$\left(\begin{array}{c|c} I_{\tau-1} & 0 \\ \hline (g_i)_i & \left(\dfrac{\partial f}{\partial x_h} + \sum_1^{\tau-1} \alpha_i \dfrac{\partial g_i}{\partial x_h}\right)_h \end{array}\right)$$

where $I_{\tau-1}$ denotes the identity matrix of rank $\tau - 1$.

It follows that C is defined by the equations:

$$R_h = \frac{\partial f}{\partial x_h} + \sum_1^{\tau-1} \alpha_i \frac{\partial g_i}{\partial x_h} = 0 , \qquad h = 1,\dots, n .$$

Since $f \in \mathfrak{m}^3$ we may suppose that $g_i(x) = x_i$ for $i = 1,\dots,n$. Then the jacobian $\left(\dfrac{\partial R_h}{\partial \alpha_k}\right)_{\substack{h=1,..,n \\ k=1,..,n}}$ coincides with the identity $(n \times n)$ matrix at $0 \in \mathbb{C}^{n+\tau}$, so that C is a non singular variety of dimension $\tau - 1$.

Also one may deduce (elimination theory) that D is defined by a single equation $h_D \in \mathbb{C}\{\alpha_1,\dots,\alpha_\tau\}$.

Before stating the properties of C and D , we shall describe in another way the deformation φ so as to get a geometric feeling of what one can expect to be true.

As we have seen, under the not restrictive hypothesis $f \in \mathfrak{m}^3$, the deformation φ has the form

$(x_1,\dots,x_n,\alpha_1,\dots,\alpha_{\tau-1}) \to (\alpha_1,\dots,\alpha_{\tau-1}, f(x) + \sum_1^{\tau-1} \alpha_i g_i(x))$ where the g_i generate $\mathfrak{m}$. This implies that the map $\sigma : \mathbb{C}^n \to \mathbb{C}^\tau$ given by

$x \rightsquigarrow (f(x), g_1(x),\dots,g_{\tau-1}(x))$

defines an embedding of some neighbourhood of $0 \in \mathbb{C}^{n+1}$ in $\mathbb{C}^\tau$.

Call X the image of σ and $y_0,\dots,y_{\tau-1}$ the coordinates on $\mathbb{C}^\tau$.

The "initial fibre" X_0 of φ can be identified with the intersection of X with the hyperplane $\{y_0 = 0\}$. The "deformed fibre" of φ over the point $(\alpha_1,\dots,\alpha_{\tau-1},\beta)$ can be identified with the intersection of X with the hyperplane $y_0 + \sum_1^{\tau-1} \alpha_i \; y_i = \beta$ (α_i , β very small).

This means that we start with a (piece of a) manifold X in $\mathbb{C}^\tau$ and an (initial) hyperplane H_0 through a point $x_0 \in X$, such that $X_0 = H_0 \cap X$ has an isolated singular point at x_0 . And we are looking at the system of hyperplane sections of X , made by hyperplanes H near to H_0 .

Let $x \in X$, $T(X,x)$ its tangent space. A point in the domain of φ can be thought of as a couple (H_ε,x)

consisting of a hyperplane (near to H_0) and a point in X such that $x \in H_\varepsilon$. Then (H_ε, x) is in the critical set C iff $H_\varepsilon \supset T(X,x)$; so the map $(H_\varepsilon, x) \rightsquigarrow x$ shows that C is locally the product $X \times \{$hyperplanes in $\mathbb{C}^\tau$ containing a fixed n-plane$\} \simeq X \times \mathbb{C}^{\tau-1}$ which shows again why C is non singular.

We recall the following :

Proposition 6 ([1], lemma 2) Let X be a non singular algebraic variety in $\mathbb{P}_N(\mathbb{C})$. Then the set of linear pencils of hyperplanes H_t in $\mathbb{P}_N$ s.th. $H_t \cap X$ is non singular or has at most one quadratic ordinary point, is a non-empty Zariski open set in the appropriate Grassmannian.

This suggests that in our case something similar is true and in fact this can be proved by local arguments, as follows : suppose first that there exists H_{ε_0} s.th. $H_{\varepsilon_0} \cap X$ has a non-degenerate quadratic singularity at some point x_0 (and perhaps other singularities). Then the set of such H_ε is dense in D because

a) almost all ("a Zariski open set") H_ε containing

$T(X,x_o)$ have the property assumed above for H_{ϵ_o} .

b) the set of x_o for which such a H_{ϵ_o} exists is a "Zariski open set" .

After that we can find **an** H_{ϵ_o} s.th. $H_{\epsilon_o} \cap X$ has exactly one singularity (at x_o) , and this is quadratic non degenerate; in fact apply Bertini's theorem (or Sard's theorem).

Again the set of such H_ϵ is dense in D .

Hence the problem is reduced to showing that at least one singular fibre has at least one quadratic non degenerate singular point.

In our case this can be proved in this way: consider the deformation φ' of X_o defined by $F(x,\lambda) = f(x) + \lambda \sum_o^n x_i^2 = 0$; since $f \in \mathfrak{m}^3$, for all $\underline{\lambda} \neq 0$ the fibre over $\underline{\lambda}$ has a non gedegenerate quadratic point. On the other hand since $\varphi : \mathbb{C}^{n+\tau} \longrightarrow \mathbb{C}^\tau$ is semiuniversal; φ' is induced by a map $\sigma : \mathbb{C} \to \mathbb{C}^\tau$, so that the fibre over $\sigma(\underline{\lambda})$ has a non degenerate quadratic point.

To discover the properties of D we look first to the behawiour of φ at points near 0 .

We suppose again that X is defined in $\mathbb{C}^{n+1} \times C$ by by $F(x,\alpha) = f(x) + \sum_1^{\tau} \alpha_i \; g_i(x) = 0$ and that $\varphi : X \to \mathbb{C}^{\tau}$ is induced by the projection $\mathbb{C}^{n+1} \times \mathbb{C}^{\tau} \to \mathbb{C}^{\tau}$.

Denote by $\mathfrak{F}$ the sheaf $\mathcal{O}_{\mathbb{C}^{n+1}\times\mathbb{C}^{\tau}} / (F, \frac{\partial F}{\partial x_0}, \dots, \frac{\partial F}{\partial x_n})$.

Since $\mathfrak{F}$ is concentrated on C and $\varphi : C \to \mathbb{C}^{\tau}$ is locally proper, the direct image sheaf $\varphi_* \mathfrak{F}$ of $\mathfrak{F}$ is a coherent $\mathcal{O}_{\mathbb{C}^{\tau}}$- sheaf - The stalk of $\varphi_* \mathfrak{F}$ at a point $\alpha \in \mathbb{C}^{\tau}$, is the direct sum of the stalks of $\mathfrak{F}$ at the (finite number of) points $x \in \varphi^{-1}(\alpha) \cap C$. On the other hand $\mathfrak{F}_x / \varphi^*(\mathfrak{m}_{\mathbb{C}^{\tau},\alpha})$ can be identified with $\mathrm{Def}((X_\alpha, x), T_1)$ where (X_α, x) denotes the germ of $\varphi^{-1}(\alpha)$ at the point x.

So $(\varphi_* \mathfrak{F})_\alpha / \mathfrak{m}_{\mathbb{C}^{\tau},\alpha}$ is $\bigoplus_{x \in \varphi^{-1}(\alpha) \cap C} \mathrm{Def}((X_\alpha, x), T_1)$.

Consider the morphism of sheaves $\lambda : \mathcal{O}_{\mathbb{C}^{\tau}} \to \varphi_* \mathfrak{F}$ induced by $(0, \dots, \underset{i}{1}, \dots, 0) \quad \frac{\partial F(x,\alpha)}{\partial \alpha_i} = g_i(x)$.

We claim that λ is surjective at the point 0, so that it will remain surjective at each point in some neighbourhood of 0.

In fact by Nakayama's lemma it suffices to prove that

$$\lambda_o : \mathbb{C}^\tau = \mathcal{O}_{\mathbb{C}^\tau,o}/\mathfrak{m}_{\mathbb{C}^\tau,o} \longrightarrow (\varphi_* \mathfrak{F})_o/\mathfrak{m}_{\mathbb{C}^\tau,o} = \mathbb{C}\{x_o,\ldots,x_n\}/(f,\frac{\partial f}{\partial x_i})$$

is surjective; since $\lambda_o(\alpha_1,\ldots,\alpha_\tau) = \sum_1^\tau \alpha_i \; g_i(x)$ and since the $g_i(x)$ are choosen so as to induce a basis of $\mathbb{C}\{x_o,\ldots,x_n\}$, it follows that λ is an isomorphism.

The surjectivity of λ implies that if α is sufficiently near to $0 \in \mathbb{C}^\tau$, the linear map $\lambda_\alpha : \mathbb{C}^\tau =$ $= \mathcal{O}^\tau_{\mathbb{C}^\tau,\alpha}/\mathfrak{m}_{\mathbb{C}^\tau,\alpha} \longrightarrow (\varphi_*\mathfrak{F})_\alpha/\mathfrak{m}_{\mathbb{C}^\tau,\alpha}$ is still surjective.

Let $\{x_1,\ldots,x_n\} = \varphi^{-1}(\alpha) \cap C$ be the singular points of X_α. Then one can easily realize that the morphism associated to φ (considered as a deformation of (X_α, x_i)) is the composition of $\lambda_\alpha : \mathbb{C}^\tau \to$ $\to \bigoplus_{j=1}^{r} \mathrm{Def}((X_\alpha, x_j), T_1)$ with the projection on $\mathrm{Def}((X_\alpha, x_i), T_1)$. This implies at once (the codomain of φ being smooth) that φ represents a versal deformation of $(X_{\varphi(x)}, x)$ for all x in some neighbourhood of 0 in X ; but one can get more :

let $(\mathbb{C}^k,\alpha) \subset (\mathbb{C}^\tau,\alpha)$ be a k plane through α in $\mathbb{C}^\tau$, such that $\lambda : T(\mathbb{C}^k, \alpha) \to (\varphi_* \mathcal{F})_\alpha$ is bijective (hence $k = \sum_1^r \tau_i$, $\tau_i = \dim \ \mathrm{Def}((X_\alpha, x_i), T_1))$.

Let $\sigma_i : (\mathbb{C}^\tau, \alpha) \to (\mathbb{C}^{\tau_i}, 0)$ be a morphism inducing the versal deformation φ of (X_α, x_i) from its semiuniversal deformation. Then $\sigma : (\sigma_1, \ldots, \sigma_r) : (\mathbb{C}^\tau, \alpha) \to \prod_1^r (\mathbb{C}^{\tau_i}, 0) = (\mathbb{C}^k, 0)$ is a surjective map, whose restriction to $(\mathbb{C}^k, \alpha)$ is a biholomorphism.

So, in this situation, one obtains the following result :

Proposition 7. The discriminant D at α can be defined by a function $h_D(\alpha) = \prod_1^r h_{D_i}(\alpha)$ where h_{D_i} depends only on τ_i variables and represents in $\mathbb{C}^{\tau_i}$ the discriminant of the semiuniversal deformation of (X_α, x_i) ; moreover for $i \neq j$, h_{D_i} and h_{D_j} depend on different variables.

This can be expressed by saying that D is an union of $D_i \times \mathbb{C}^{\tau - \tau_i}$ transversal each other in the best possible way .

As an example, if X_α has two singular points x_1 ,

x_2 , x_1 of the type $\sum_0^n x_i^2 = 0$ and x_2 of the type $x_0^3 + \sum_1^n x_i^2 = 0$, then $\tau \geq 3$ and one can find local coordinates at $\alpha \in \mathbb{C}^\tau$ so that $h_D = \alpha_1 \cdot (\alpha_2^2 - \alpha_3^3)$, as one sees by computing the discriminant of the semiuniversal deformation of such a kind of singularities. However, as we shall see, it must be possible to finde some other information on the behaviour of D : one can prove that actually one must have in such a situation $\tau \geq 4$.

Examples : a) $X_0 : \sum_0^n x_i^2 = 0$. Its semiuniversal deformation is given by $X \subset \mathbb{C}^{n+1} \times \mathbb{C}$ defined by $\sum_0^n x_i^2 = \beta$. Hence its discriminant defined by $\beta = 0$ is just a (reduced) point

b) $X_0 : x_0^3 + \sum_1^n x_i^2 = 0$. Its semiuniversal deformation is defined in $\mathbb{C}^{n+1} \times \mathbb{C}^2$ by $x_0^3 + \sum_1^n x_i^2 - \alpha x = \beta$; the discriminant being defined by the equation $4\alpha^3 = 27\beta^2$, hence essentially $\alpha^3 = \beta^2$, is an ordinary cusp.

We are now ready to investigate the properties of C ,

D and how they are related to the singularities of the fibres of φ .

Proposition 8.

i) C is smooth

ii) D is reduced

iii) $\varphi : C \rightarrow D$ is the normalization of D , in particular D is irreducible.

iv) the multiplicity of D at a point α is $\sum_{1}^{r} \mu(x_i)$ where $x_1,\ldots,x_r$ are the singularities of X and $\mu(x_i)$ denotes the dimension over $\mathbb{C}$ of $\mathbb{C}\{x_0,\ldots,x_n\} / \left(\frac{\partial g}{\partial x_0},\ldots,\frac{\partial g}{\partial x_n}\right)$, where $g \in \mathbb{C}\{x_0,\ldots,x_n\}$ is a local equation of (X_α , x_i) .

v) D is smooth at a point α iff $X_\alpha = \varphi^{-1}(\alpha)$ has only one singularity, and this is of the type $\sum_{0}^{n} x_i^2 = 0$

vi) There exists a closed analytic set $\Lambda \subset D$ of codimension ≥ 3 in $\mathbb{C}^{\tau}$ such that the only singularities of $D - \Lambda$ are of the type

a) $\alpha_1 \cdot \alpha_2 = 0$

b) $\alpha_1^2 = \alpha_2^3$

Moreover at a point $\alpha \in \mathbb{C}^{\tau}$, D has a local equation of the type a) iff X_{α} has exactly two singularities, both of the type $\sum_{0}^{n} x_i^2 = 0$; D has a local equation of the type b) iff. X_{α} has just one singular point, and this is of the type $x_0^3 + \sum_{1}^{n} x_i^2 = 0$.

Proof. i) has been just proved. By considering the deformation $f(x) + \lambda \sum_{0}^{n} x_i^2 = 0$, one gets that some fibre of φ has at least one double point (remark that we have supposed $f \in \mathfrak{m}^3$).

From proposition 7. we get that some fibre has only one singularity, this being that of an ordinary quadratic point; hence D is smooth somewhere; since D is defined by a single equation this proves ii) and it implies that $\varphi: C \to D$ is of degree one; hence we get iii) because C is smooth .

By proposition 7. to prove iv) it suffices to prove it for $\alpha = 0$. Let $\gamma : D \to \mathbb{C}^{\tau-1}$ be the projection on the variables $\alpha_1, \ldots, \alpha_{\tau-1}$. Then $\gamma \circ \varphi : C — \mathbb{C}^{\tau-1}$ is a finite map whose fibre over 0 is one point with structure sheaf $\mathbb{C}\{x_0,\dots,x_n\}\Big/\left(\frac{\partial f}{\partial x_0},\dots,\frac{\partial f}{\partial x_n}\right)$. Hence $\gamma\circ\varphi$ has degree $\mu(0)$ and since $\varphi : C \to D$ has degree one, this implies that $\gamma : D \to \mathbb{C}^{\tau-1}$ has degree $\mu(0)$, so that the multiplicity $m_0(D)$ of D at 0 is less or equal to $\mu(0)$. On the other hand if $\tilde{\gamma}$ is any other projection of D on a $\mathbb{C}^{\tau-1}$, the composition $\tilde{\gamma}\circ\varphi : C \to \mathbb{C}^{\tau-1}$ has degree $\geq \mu(0)$, so that $m_0(D) = \mu(0)$.

Now v) is nothing else than the Morse lemma : if D is smooth at α then D is locally irreducible, so that X_α has only one singularity; let g be a local equation for that singularity. Since D has multiplicity one, it follows that $\mathbb{C}\{x_0,\dots,x_n\}\Big/\left(\frac{\partial g}{\partial x_0},\dots,\frac{\partial}{\partial x_n}\right)$ has dimension one, so that $\frac{\partial}{\partial x_0},\dots,\frac{\partial}{\partial x_n}$ generate the maximal ideal in $\mathbb{C}\{x_0,\dots,x_n\}$ and hence by Morse lemma can be supposed of the form $(x) = \sum_0^n x_i^2$.

Now we consider vi) ; let $\Lambda = \{\alpha\in\mathbb{C}^\tau \mid D$ has multiplicity ≥ 3 at $\alpha\}$. We claim that Λ has codimension at least three in $\mathbb{C}^\tau$.

In fact let $H : C \to \mathbb{C}$ be the function defined by

$$H(x,\alpha) = \det\left(\frac{\partial^2(f(x)+\sum_1^{\tau-1}\alpha_i\ {}_i(x)}{\partial x_i\ \partial x_j}\right), \quad E \text{ the space}$$

defined by $\{H = 0\}$, $S=\left\{(x,\alpha)\in E \,\middle|\, \frac{\partial H}{\partial x_i} = 0,\ i=0,\dots,n\right\}$.

Denote by $X_{x,\alpha}$ the germ at (x,α) of $\varphi^{-1}\varphi(x,\alpha)$.

One has that

a) $(x,\alpha) \in E$ iff $X_{x,\alpha}$ is not an ordinary quadratic point

b) $(x,\alpha) \in S$ iff $X_{x,\alpha}$ is not a singularity of the type

$$x_0^\sigma + \sum_1^n x_i^2 = 0, \text{ with } \sigma = 2 \text{ or } 3.$$

a) and b) are proved with the aid of Morse lemma:

a) is clear; for b), since the hypothesis that for some i, $\frac{\partial H}{\partial x_i} \neq 0$, implies that the Hessian of the function g defining $X_{x,\alpha}$ has characteristic n, applying Morse lemma one may suppose $g = x_0^\sigma + \sum_1^n x_i^2$, and from that b) is easily deduced.

We have seen that the generic $(x,\alpha) \in C$ is such that $X_{x,\alpha}$ is an ordinary double point : by a) this means that $H \not\equiv 0$ so that $\dim E = \dim C - 1 = \tau - 2$.

To prove that $\dim S \leq \tau - 3$ it suffices by b) to

show that in the semiuniversal deformation of any singularity $X_{x,\alpha}$ that is not an ordinary quadratic point, there are fibres with a singularity of the type $x_0^3 + \sum_1^n x_i^2 = 0$, i.e. ordinary cusps ; in fact if $g(x)$ defines $X_{x,\alpha}$, let $g(x) = \tilde{g}(x_0,\dots,x_m) + \sum_{m+1}^n x_i^2$, where the hessian of $\tilde{g}$ vanishes at x.

Then $\tilde{g}(x_0,\dots,x_m) + \sum_{m+1}^n x_i^2 + \lambda(x_0^3 + \sum_1^m x_i^2) = 0$

defines a deformation of $X_{x,\alpha}$ which has "definitively" an ordinary cusp for $\lambda \to 0$.

Now since $\varphi : C \to \mathbb{C}^\tau$ is finite, $\dim \varphi(S) = \dim S \leq \tau - 3$.

Let now $\alpha \in \Lambda - \varphi(S)$; then X_α has only singularities of the type : ordinary quadratic point or cusp; by proposition 7. we deduce now that $\dim_\alpha \Lambda \leq \tau - 3$.

Denote by $(\alpha_1,\dots,\alpha_{\tau-1},\beta)$ the coordinates on $\mathbb{C}^\tau$, where $f(x) + \sum_1^{\tau-1} \alpha_i\, g_i(x) = \beta$ is the equation in $\mathbb{C}^{n+1} \times \mathbb{C}^\tau$ defining the domain X of the semiuniversal deformation φ of X_0.

We have seen that D is intersected by the β-axis

with minimal multiplicity $\mu = \dim_{\mathbb{C}} \mathbb{C}\{x_o,\ldots,x_n\} / \left(\frac{\partial f}{\partial x_i}\right)$,

so that, by the Weierstrass preparation theorem, D is defined by an equation of the type $\beta^{\mu} + a_1(\alpha)\beta^{\mu-1} + \ldots$

$\ldots + a_{\mu}(\alpha) = 0$, where

$a_i(\alpha) \in \mathfrak{m}$ = maximal ideal in $\mathbb{C}\{\alpha_1,\ldots,\alpha_{\tau-1}\}$, $i=1,\ldots,\tau-1$

We shall show now that in some sense D is "parallel" to the hyperplane in $\mathbb{C}^{\tau}$ defined by $\beta = 0$. It will follows in particular that $a_i(\alpha) \in \mathfrak{m}^{i+1}$, $i=1,\ldots,\tau-1$.

We recall first the concept of tangent cone of a germ (X,x).

Let X be an analytic subset in $\mathbb{C}^n$, $x \in X$. The tangent cone $C(X,x)$ of X at x is defined (see [19] Whitney) by $C(X,x) =$

$= \{ v \in \mathbb{C}^n \mid$ there exists sequences (x_n) in X and (λ_n) in $\mathbb{C}$ s.th. $\lambda_n(x_n - x)$ converges to $v \}$.

<u>Lemma</u> 4.([19]) Let I be the ideal in $\mathbb{C}\{x_1,\ldots,x_n\}$ defining X at $x = 0 \in \mathbb{C}^n$; for $f \in I$, denote by $\tilde{f}$ the homogeneous part of lower degree of f. Then

$I_o = \{ \tilde{f} \mid f \in I \}$ is an (homogeneous) ideal in $\mathbb{C}[x_1,..,x_n]$ whose locus of zeros in $\mathbb{C}^n$ coincides with $C(X,x)$ (in general I_o is not a reduced ideal: for example if X is defined by $x^2 - y^3 = 0$ in $\mathbb{C}^2$, then $I_o = x^2 \cdot \mathbb{C}[x,y]$).

<u>Proposition</u> i) the tangent cone $C(D,0)$ is defined by $\beta = 0$ and it coincide with the image of $d\varphi(0) : T(X,0) \rightarrow T(\mathbb{C}^\tau, 0)$.

ii) D is defined by an equation of the type $\beta^\mu + a_1(\alpha)\beta^{\mu-1} + \ldots + a_\mu(\alpha) = 0$ where for $i = 1,\ldots,\mu$, $a_i(\alpha)$ vanishes at 0 of order at least $i+1$.

iii) For $d \in D$, $C(D,d)$ consists of r hyperplanes in generic position in $\mathbb{C}^\tau$, where r is the number of singularities of $\varphi^{-1}(d)$. Moreover for $d \rightarrow 0$, these hyperplanes converge to the hyperplane $\beta = 0$.

<u>Proof</u>. Since we may suppose that f vanishes at $0 \in \mathbb{C}^{n+1}$ of order at least three, we may choose $g_1 = - x_o$, $\ldots, g_{n+1} = - x_n$ so that φ has the following form :

$$(\alpha_1, \ldots, \alpha_{\tau-1}, x_o, \ldots, x_n) \to (\alpha_1, \ldots, \alpha_{\tau-1}, f(x) -$$

$$- \sum_1^{n+1} \alpha_i x_{i-1} + \sum_{n+2}^{\tau-1} \alpha_j g_j(x)) .$$

Where each g_i vanishes of order at least two.

It follows that the critical set C is defined by the equations :

$$\alpha_i = \frac{\partial f}{\partial x_{i-1}} + \sum_{n+2}^{\tau-1} \alpha_j \frac{\partial g_j}{\partial x_{i-1}} , \quad i = 1, \ldots, n+1 .$$

In particular $(x_o, \ldots, x_n, \alpha_{n+2}, \ldots, \alpha_{\tau-1})$ define a system of coordinates on C . Hence the map $\varphi : C \to \mathbb{C}^\tau$ has the form

$$(*) \begin{cases} \alpha_i = \dfrac{\partial f}{\partial x_{i-1}} + \displaystyle\sum_{n+2}^{\tau-1} \alpha_j \dfrac{\partial g_j}{\partial x_{i-1}} , & i = 1, \ldots, n+1 \\ \alpha_i = \alpha_i & , \quad i = n+2, \ldots, \tau - 1 \\ \beta = f(x) - \displaystyle\sum_1^{n+1} x_{i-1} \cdot \left(\dfrac{\partial f}{\partial x_{i-1}} + \displaystyle\sum_{n+2}^{\tau-1} \alpha_j \dfrac{\partial g_j}{\partial x_{i-1}} \right) + & \\ \quad + \displaystyle\sum_{n+2}^{\tau-1} \alpha_j \ g_j(x) . & \end{cases}$$

Suppose λ varies in a sequence in $\mathbb{C}$ and that

(x,α) varies in a sequence in C s.th. $\lambda\cdot\varphi(x,\alpha)$ has a limit in $\mathbb{C}^{\tau}$ (so, that limit is a point in $C(D,0)$). By (*) we get that for $i = n+2,\ldots,\tau-1$, $\lambda\alpha_i$ has a finite limit, so that $\lambda\alpha_i\cdot g_i$ converges to zero. Since for $i = 1,\ldots,n+1$, $\lambda\cdot\alpha_i$ has a finite limit; it follows that $\lambda\frac{\partial f}{\partial x_{i-1}}$ has a finite limit. Since (for example by an inequality of Lojasiewicz [ ] p. 92) $|f(x)| \leq \sum_{0}^{n}\left|\frac{\partial f}{\partial x_i}\cdot x_i\right|$

in some neighbourhood of zero, it follows that $\lambda\cdot\beta$ has limit zero.

This proves that $C(D,0)$ is contained in the hyperplane $\beta = 0$; since $C(D,0)$ is an ,
it follows that $C(D,0)$ is defined by $\beta = 0$.

To prove i) it remains only to remark that the image of $d\varphi$ is exactly the hyperplane $\beta = 0$ and this is clear .

From i) and the proposition 7 , we deduce iii) by continuity. Moreover, since we have just noticed that D is defined by an equation of the type $\beta^{\mu} + a_1(\alpha)\beta^{\mu-1} + \ldots$
$\ldots + a_{\mu}(\alpha) = 0$, ii) is obviously deduced from i) .

7. The fibration associated to a deformation.

Let $\pi: (\mathbb{C}^N,0) \rightarrow (\mathbb{C}^n,0)$ be a flat morphism whose fibre $(X_o,0) = \pi^{-1}(0)$ is a hypersurface with an isolated singularity at 0 (for example π is the semiuniversal deformation of the hypersurface singularity $(X_o,0)$) .

Denote by $(\Delta,0)$ the discriminant of π .

We are interested in the topological description of that deformation; to do this one needs to choose a map between neighborhoods of $0 \in \mathbb{C}^N$ and $0 \in \mathbb{C}^n$ that represents the map π between germs of spaces; moreover this choice must be, in some way, intrinsically associated to π .

One works in the following way: first of all fix a disk D around 0 in $\mathbb{C}^N$, very small compared with the fibre $\pi^{-1}(0)$; this means (see Milnor [14]) that D and every other disk around o smaller than D has a boundary (sphere) which is transversal to $\pi^{-1}(0)$. After that we choose a ball B around 0 in $\mathbb{C}^n$ so small that the following two conditions be verified

1) π restricted to $\partial D \cap \pi^{-1}(B)$ has no critical points.

2) The boundary of B and that of every other ball around o smaller than B, are "transversal" to Δ.

Since Δ is not (in general) a manifold outside 0 the meaning of the word "transversal" in condition 2 must be specified. This is done by the use of a stratification of Δ (i.e. by a partition of Δ into manifolds, which satisfies certain conditions on the angles between their tangent spaces, see Whitney [19]) and by requiring the transversality with each stratum (manifold of the partition) of the stratification.

One can show (see [ ]) with the aid of isotopy theorems that if D and B are choosen in this way, then

a) $\pi : D \cap \pi^{-1}(B) \to B$ is a smooth proper map

b) $\pi : D \cap \pi^{-1}(B-\Delta) \to B - \Delta$ is a differentiable fibre bundle, whose fibre M is a compact differentiable $2r$ - dimensional manifold with boundary, where $r = N-n$; moreover, since π has no critical points on $\pi^{-1}(B) \cap D$, one may suppose that the group of the bundle is made with diffeomorphisms of M which are the identity on ∂M.

c) If D', B' are choosen so as to satisfy conditions 1), 2), then $\pi : D \cap \pi^{-1}(B) \to B$ is homeomorphic with $\pi : D' \cap \pi^{-1}(B') \to B'$.

So the fibre bundle $\pi : D \cap \pi^{-1}(B-\Delta) \to B-\Delta$ is (up to isomorphisms) intrinsically associated with $\pi : (\mathbb{C}^N,0) \longrightarrow (\mathbb{C}^n,0)$; topological properties of that fibre bundle (or of the map $\pi : D \cap \pi^{-1}(B) \to B$) will be called topological properties of π.

We shall be mainly interested into homological computations about the fibre bundle π ; that means to compute the homology $H_*(M,\mathbb{Z})$ of the fibre $M = \pi^{-1}(x) \cap D$, $x \in B - \Delta$, and then the representation

$$\sigma : \pi_1(B-\Delta, x) \to \mathrm{Aut}(H_*(M,\mathbb{Z})) .$$

Milnor [14] considers this situation for $n = 1$; here, also the case $n = 1$, will be studied through the geometric description of its semiuniversal deformation, looking expecially at the corresponding discriminant.

For that we shall need a complete description only for the map $\pi : (\mathbb{C}^n,0) \to (\mathbb{C},0)$ defined by

$\pi(z_1,\ldots,z_n) = \sum_1^n z_i^2$.

Let $x = (x_1,\ldots,x_n)$, $y = (y_1,\ldots,y_n)$ denote the real and imaginary part of $z = (z_1,\ldots,z_n)$, i.e. $z_\alpha = x_\alpha + i\, y_\alpha$, $\alpha = 1,\ldots,n$, so that $x,y \in \mathbb{R}^n$ and $z = x + i\, y$. In this way $\mathbb{C}^n$ is identified with $\mathbb{R}^n \oplus \mathbb{R}^n$ and π can be written as $\pi(x,y) = \|x\|^2 - \|y\|^2 + 2i <x,y>$ where $< , >$ denotes the standard scalar product on $\mathbb{R}^n$. We make the following remarks :

a) the fibre $X_0 = \pi^{-1}(0)$ has an isolated singularity at $0 \in \mathbb{C}^n$; all the others fibres of π are diffeomorphic each other. In fact for $\lambda \in \mathbb{C} - \{0\}$, define a diffeomorphism from $X_1 = \pi^{-1}(1)$ to $X_\lambda = \pi^{-1}(\lambda)$ in this way : let $\underline{\lambda}$ be a square root of λ , i.e. $\underline{\lambda}^2 = \lambda$; then $(z_1,\ldots,z_n) \longmapsto (\underline{\lambda}\, z_1,\ldots,\underline{\lambda}\, z_n)$ maps X_1 onto X_λ and it defines a diffeomorphism.

b) The fibre X_1 is diffeomorphic with the tangent bundle $T(S^{n-1})$ of the sphere S^{n-1} . Moreover the intersection $X_1 \cap D_n(\varepsilon)$ can be identified with the set of $v \in T(S^{n-1})$ whose lenght is less or equal to $\sqrt{\frac{\varepsilon-1}{2}}$. The diffeomorphism written above between X_1 and the X_λ

are not good for our purpose. In fact we want diffeomorphism which "induce the identity on the boundary" . The reason is the following.

Suppose to have a family of compact complex space $(X_t)_{t\in\mathbb{C}}$ such that X_t is non singular for $t \neq 0$ and X_0 has just one singularity x_0 . Then one is interested to know how the cycles of say X_1 vary when t turns once around the origin $0 \in \mathbb{C}$. If $D(x_0,\varepsilon)$ is a small disk around the singular point of X_0 , one knows that the family $(X_t - D(x_0,\varepsilon))_{|t|\leq\eta}$ has the structure of a product $X_0 - D(x_0,\varepsilon) \times \{|t| \leq \eta\}$ if η is sufficiently small. It follows that if γ is a cycle in X_η which does not "intersect" $X_\eta \cap D(x_0,\varepsilon)$, i.e. it can be represented by a chain in $X_\eta - D(x_0,\varepsilon)$, then γ will be an "invariant cycle" , that means one can follow the family $(X_t)_{|t|=\eta}$ in such a way that the chain representing γ (and so γ itself) is transformed into itself. Moreover every cycle can be represented as the "sum" of a chain in $X_\eta \cap D(x_0,\varepsilon)$ and a chain in $\overline{X_\eta - D(x_0,\varepsilon)}$ whose boundaries are in $D(x_0,\varepsilon)$; since only the first of these two chain will be changed as t rounds around $0 \in \mathbb{C}$, it follows the importance to study the family $(X_t \cap D(x_0, \varepsilon))_{|t|=\eta}$ but taking the "boundaries" of this family fixed (that means on the boundaries we have just given a trivialization induced by trivializating the family outside $D(x_0, \varepsilon)$).

In the example we are considering, this trivialization can be made explicitly through the identification of $X_t \cap D(x_0, \varepsilon)$ with a tubular neighbourhood of the zero section in the tangent bundle of the sphere S^{n-1}.

Let $\varepsilon > 1$; then $X_1 \cap D(0, \varepsilon)$ is defined by $\|x\|^2 - \|y\|^2 = 1$, $\langle x, y \rangle = 0$, $\|x\|^2 + \|y\|^2 \leqslant \varepsilon$ and it is identified with the space $N = \{(x', y') \in \mathbb{R}^n \times \mathbb{R}^n \mid \|x'\| = 1, \|y'\| \leq \sqrt{\frac{\varepsilon - 1}{2}}, \langle x', y' \rangle = 0\}$ by the map $\sigma : (x, y) \longmapsto (\frac{x}{\|x\|}, y)$. Let $\alpha = \sqrt{\frac{\varepsilon - 1}{2}}$ and $\rho : [0, +\infty[\longrightarrow \mathbb{R}$ a differentiable function s.th. $\rho(t) = 0$ for $0 \leqslant t \leqslant \frac{\alpha}{3}$, $\frac{d\rho}{dt} > 0$ for $\frac{\alpha}{3} < t < \frac{2}{3}\alpha$, and $\rho(t) = 1$ for $t \geqslant \frac{2}{3}\alpha$.

For $\varphi \in [0, 2\pi]$, define a diffeomorphism $\tau(\varphi) : N \longrightarrow N$ in the following way: $\tau(\varphi)(x', y')$ is obtained by a rotation in the plane in $\mathbb{R}^n$ that contains x', y' of an angle $\theta = \rho(\|y'\|) \cdot \varphi/2$, applied to the vectors x', y'. The analytic form of $\tau(\varphi)$ is the following:

$$(x', y') \rightarrow (\cos\theta \cdot x' + \sin\theta \cdot \frac{y'}{\|y'\|}, -\|y'\| \cdot \sin\theta \cdot x' + \cos\theta \cdot y')$$

Now we are able to write down good trivializating dif

feomorphisms $X_1 \cap D(0,\varepsilon) \to X_\lambda \cap D(0,\varepsilon)$ for $\lambda = e^{i\varphi}$ by the composition $X_1 \cap D(0,\varepsilon) \xrightarrow{\sigma} N \xrightarrow{I\varphi} N \xrightarrow{\sigma^{-1}} X_1 \cap D(0,\varepsilon) \xrightarrow{\gamma} X_\lambda \cap D(0,\varepsilon)$ where γ is induced by $(z_1,\dots,z_n) \to (e^{i\varphi/2}\cdot z_1,\dots,e^{i\varphi/2}\cdot z_n)$. We are interested in the diffeomorphism corresponding to $\varphi = 2\pi$. If we "read" it through σ as an automorphism of N we find it has the following analytic form:

(*) $h:(x,y) \to (-(\cos\theta \cdot x + \sin\theta \frac{y}{\|y\|}), \|y\|\cdot\sin\theta\cdot x - \cos\theta\cdot y)$

where $\theta = \rho(\|y\|)\cdot\pi$.

Remark that h, in a neighbourhood of the zero section in N (i.e. $\|y\|$ small), acts as the antipodal map: $(x,y) \to (-x,y)$ and in a neighbourhood of the boundary of N it induces the identity.

We want to compute the action h_* of h on $H_*(N \bmod \partial N, \mathbb{Z})$. Since h is the identity on ∂N, it follows that $h_* = id + v$ where id denotes the identity on $H_*(N \bmod \partial N, \mathbb{Z})$ and v is an homomorphism $H_*(N \bmod \partial N, \mathbb{Z}) \to H_*(N, \mathbb{Z})$ (which is canonically mapped into $H_*(N \bmod \partial N, \mathbb{Z})$).

Since $H_*(N,\mathbb{Z}) \simeq H_*(S^{n-1},\mathbb{Z})$, we have to study only $v: H_{n-1}(N \bmod \partial N, \mathbb{Z}) \to H_{n-1}(N,\mathbb{Z}) \simeq e.\mathbb{Z}$, where e is the cycle (uniquely determined up to the sign) induced by the fundamental class of the zero section S^{n-1} in N.

To compute v, let $z \in H_{n-1}(N \bmod \partial N, \mathbb{Z})$. Then one can represents z by a chain ζ in N whose boundary lies in ∂N and s.th. at each intersection point of the support $|\zeta|$ of ζ with the zero section S^{n-1} of N, the intersection is transversal (for example $|\zeta|$ coincides locally with the fibre of $N \to S^{n-1}$); then at each point of $|\zeta| \cap S^{n-1}$ one can associate a number ± 1, according to whether the orientation of $|\zeta|$ and the choosen orientation of S^{n-1} induce the canonical orientation of N or not (remark that N is canonically oriented, because of its complex structure). Denote by $< , > : H_{n-1}(N \bmod \partial N, \mathbb{Z}) \to H_{n-1}(N, \mathbb{Z})$ the intersection product and call Δ the element in $H_{n-1}(N \bmod \partial N, \mathbb{Z})$ induced by a (triangulation of a) fibre of $N \to S^{n-1}$, oriented in such a way that $< \Delta , e > = +1$.

Since h is the identity near N, we get the following formula:

$$v(\zeta) = < \zeta , e > \cdot v(\Delta)$$

which reduces the computation of v to that of $v(\Delta)$. Now, if the support $|\Delta|$ of Δ is the fibre of $N \to S^{n-1}$ over the point $(1,0,\dots,0)$ (and hence it is parametrized by $y=(y_2,\dots,y_n) \in \mathbb{R}^{n-1}$, $\|y\|^2 < \frac{\varepsilon-1}{2}$), formula (*) shows that the composition of h with the projection $N \to S^{n-1}$ gives a map $|\Delta| — S^{n-1}$ whose analytic form is:

$x' = (-\cos(\rho(\|y\|)\cdot\pi), \sin(\rho(\|y\|)\cdot\pi)\cdot\frac{y_2}{\|y\|}, \ldots, \sin(\rho(\|y\|)\cdot\pi)\cdot\frac{y_n}{\|y\|})$

By this map the disk $\|y\| \leq \frac{\alpha}{3}$ is mapped into the point $(-1,0,\ldots,0)$, and the corona $\frac{1}{3}\alpha < \|y\| < \frac{2}{3}\alpha$ is mapped homeomorphically onto $S^{n-1} - \{(1,0,\ldots 0), (-1,0,\ldots,0)\}$.

It follows that $v(\Delta) = \pm e$, where we have to choose the right sign. Choose on S^{n-1} the orientation given at the point $(1,0,\ldots,0)$ by the co-ordinates $(x_2,\ldots,x_n)$; since the orientation induced on N by the complex structure is given by $(x_2,y_2,\ldots,x_n,y_n)$ which is obtained by $\frac{(n-1)\cdot(n-2)}{2}$ transpositions, it follows that the right orientation on $|\Delta|$ is $(-1)^{\frac{(n-1)\cdot(n-2)}{2}}$ the one given by $(y_2,\ldots,y_n)$.

Now remark that the map $|\Delta| \longrightarrow S^{n-1}$ changes these orientations, and deduce the Picard-Lefschetz formula

$$v(z) = (-1)^{\frac{n\cdot(n+1)}{2}} \langle z,e\rangle \cdot e$$

We need also to know $\langle e,e\rangle$. Remark that since N is homeomorphic with the tangent bundle to S^{n-1}, $\langle e,e\rangle$ coincides with the Euler characteristic of S^{n-1} if N is oriented as the normal bundle to S^{n-1}; since, as we have seen, this is $\frac{(n-1)\cdot(n-2)}{2}$ the orientation induced on N by its complex structure, we get

$$< e,e > = \begin{cases} 0 & n \text{ even} \\ (-1)^{\frac{n-1}{2}} & n \text{ odd} \end{cases}$$

We recall briefly also the classical description (see [13]) of a family of curves which acquire a node, that means the case of a function $f(x,y) = x^2 - y^2$. One considers the non singular fibre $\Gamma_\lambda : x^2-y^2 = \lambda$, $\lambda \in \mathbb{C}-\{0\}$; one has a double covering of $\mathbb{C}$, defined by $y = \sqrt{x^2-\lambda}$, branched over the two square roots of λ.

One considers Γ_1 as the initial fibre of the family; then a closed curve in $\mathbb{C}$ containing in its interior 1 and -1 , can be lifted in Γ_1 into two curves, one of which is choosen as the vanishing cycle e at $\lambda=0$. When λ turns around the origin, the two branching points are changed one into the other and one can easily visualize the deformation of Γ_1 into itself and the Picard-Lefschetz formula.

F. Lazzeri

8 Picard - Lefschetz theory. ([5])

Let $\varphi : (\mathbb{C}^{n+h}, 0) \longrightarrow (\mathbb{C}^h, 0)$ be the semiuniversal deformation of an isolated singularity of hypersurface type (X, x), and denote by Δ the discriminant locus of φ. We have seen in 7. how if U is a small neighbourhood of $0 \in \mathbb{C}^H$, to each loop γ in $U - \Delta$ there is associated an element $\gamma_* \in \operatorname{Aut} H_*(M, \mathbb{Z})$, where M is a 2n-dimensional compact manifold with boundary, the Picard-Lefschetz theory shows how from informations about the geometry of Δ, one can deduce informations about the map $\gamma \longrightarrow \gamma_*$.

We need first to show a way to find generators and relations for the group $\pi_1(U - \Delta, x)$ where U is a small and good (relatively to Δ) neighborhood of 0, $x \in U - \Delta$.

This will be done by an ancient method in algebraic geometry (the proof we shall give is essentially that of Van Kampen [18]). First, to avoid the use of "U small and good" let us define $\pi_1(\mathbb{C}^h - \Delta, 0)$; choose a continuous arc $\gamma: [0, 1[\rightarrow \mathbb{C}^r$ with $\gamma(0) = 0$, $\gamma(]0, 1[) \subset \mathbb{C}^r - \Delta$. If U is any neighborhood of 0 in $\mathbb{C}^r$ (where Δ is defined) let $t \in]0, 1[$ be such that $\gamma(]0, t[) \subset U - \Delta$. Then the group $\pi_1(U - \Delta, \gamma) = \pi_1(U - \Delta, \gamma(t))$ is well defined and it depends only on U (once one has fixed). If $V \subset U$ one has a morphism $\pi_1(V - \Delta, \gamma) \to \pi_1(U - \Delta, \gamma)$; the limit of this system is denoted by $\pi_1(\mathbb{C}^r - \Delta, \gamma)$. If $\tilde{\gamma}$ is another arc issued from 0, one has an isomorphism $\pi_1(\mathbb{C}^r - \Delta, \gamma) \to \pi_1(\mathbb{C}^r - \Delta, \tilde{\gamma})$ determined up to inner automorphism; all of this defines $\pi_1(\mathbb{C}^r - \Delta, 0)$.

This definition works for Δ a closed set in a topological space X which is not disconnected locally at one point by Δ.

Now let Δ be a germ of an analytic set at $0 \in \mathbb{C}^r$. It is easy to see that one may suppose that Δ is of pure codimension 1.

Let $w^m + a_1(z)\, w^{m-1} + \dots + a_m(z) = 0$ be a local equation for Δ, where $(w, z_1, \dots, z_{r-1})$ are local coordinates at $0 \in \mathbb{C}^r$, and $a_i(0) = 0$, $i = 1, \dots, m$. Consider a nice stratification of Δ, say a stratification verifying Whitney's conditions. By an usual argument one can see that there exists $\varepsilon_0 > 0$ s.th. for $0 < \varepsilon < \varepsilon_0$ the hypersurface $\|z\| = \varepsilon$ is transversal to Δ (i.e. to each stratum of Δ). It follows that, if $U_\varepsilon = \{(w,z) \mid \|z\| < \varepsilon\}$ then for $0 < \varepsilon' < \varepsilon < \varepsilon_0$ the inclusion $U_{\varepsilon'} - \Delta \longrightarrow U_\varepsilon - \Delta$ is a homotopy equivalence. Moreover let $\eta(\varepsilon) = \sup_{U_\varepsilon \cap \Delta} |w|$,

$U_{\varepsilon,\eta(\varepsilon)} = \{ (w,z) \in U_\varepsilon \mid |w| < \eta(\varepsilon) \}$. Then the inclusion $U_{\varepsilon,\eta(\varepsilon)} - \Delta \longrightarrow U_\varepsilon - \Delta$ is a homotopy equivalence. Since $a_i(0) = 0$ for $i = 1,\ldots,m$, one has that $\eta(\varepsilon) \to 0$ as $\varepsilon \longrightarrow 0$, so that $U_{\varepsilon,\eta(\varepsilon)}$ is an arbitrary small neighbourhood of 0 in $\mathbb{C}^r$. In particular if $0 < \varepsilon < \varepsilon_0$ and $p \in U_\varepsilon - \Delta$, then $\pi_1(\mathbb{C}^r - \Delta, 0) \longrightarrow \pi_1(U_\varepsilon - \Delta , p)$ is an isomorphism.

<u>NOTATIONS</u>. $U = U_\varepsilon$; $V = U \cap \{w = 0\}$; $\varphi : U \to V$ the projection. Fix $|w_0| \gg \varepsilon$. For $z \in V$, L_z is the straight line $\varphi^{-1}(z)$ and $p_z = (w_0,z) \in L_{z_0}$.

Finally let Γ denote the discriminant of $\varphi : U \cap \Delta \to V$, $\tilde{\Gamma} = \varphi^{-1}(\Gamma)$.

Suppose $z_0 \in V - \Gamma$, then one has a diagram:

$$\begin{array}{ccccccccc} 0 \to & \pi_1(L_{z_0} - \Delta) & \xrightarrow{j} & \pi_1(U - (\Delta \cup \tilde{\Gamma})) & \underset{\gamma}{\overset{\beta}{\rightleftarrows}} & \pi_1(V - \Gamma) & \to 0 \\ & & & \downarrow \alpha & & & \\ & & & \pi_1(U - \Delta) & & & \end{array}$$

where the base point is always p_{z_0} , j and α are induced by inclusions, β by φ and γ by the map $z \mapsto (w_0, z)$. Remark that $\varphi : U - (\Delta \cup \tilde{\Gamma}) \to V - \Gamma$ is a fibre bundle with fibre $L_{z_0} - \Delta$ and that $z \to (w_0, z)$ induces a cross sec tion of such a bundle. So from the homotopy sequence of a fibre bundle we get

1) the horizontal line is exact and $\beta \circ \gamma$ is the identity on $\pi_1(V - \Gamma)$.

Moreover one has obviously

2) α is surjective

3) $\alpha \circ \gamma$ is the null homomorphism

Consider the sequence $0 \rightarrow \pi_1(V - \Gamma) \xrightarrow{\gamma} \pi_1(U - (\Delta \cup \tilde{\Gamma})) \rightarrow \pi_1(U - \Delta) \rightarrow 0$.

This must not be exact at $\pi_1(U - (\Delta \cup \tilde{\Gamma}))$. Nevertheless one has

4) $\ker \alpha$ is generated by the conjugated of $\operatorname{Im} \gamma$.

<u>PROOF</u>. Obviously for $v \in \operatorname{Im} \gamma$ and $b \in \pi_1(U - (\Delta \cup \tilde{\Gamma}))$ one has $b^{-1}vb \in \ker \alpha$. Let $b \in \ker \alpha$. Then $b = \partial c$ with $c \in \pi_2(U - \Delta, \; U - (\Delta \cup \tilde{\Gamma}))$. Let us represent c by a map $\hat{c} : [0, 1] \times [0, 1] \rightarrow U - \Delta$ which is transversal to $\tilde{\Gamma}$. Then $\hat{c}^{-1}(\tilde{\Gamma})$ is a finite set of points, let say $P_1, \ldots, P_s$. The following picture shows that b is equivalent in $\pi_1(U - (\Delta \cup \tilde{\Gamma}))$ to a product $w_1 \ldots w_s$ of simple loops around $\tilde{\Gamma}$:

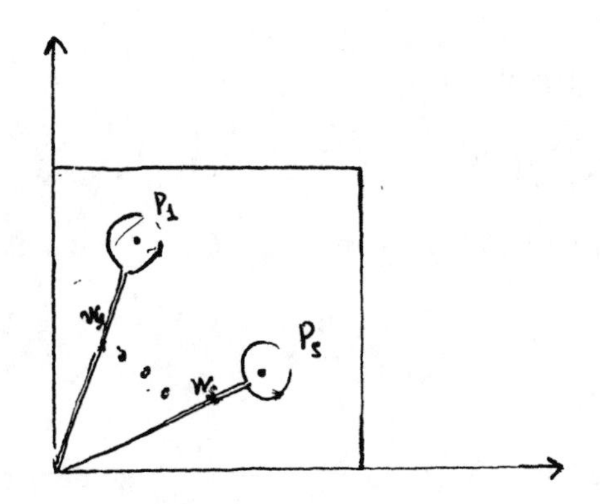

w_i simple means that it is composed of an arc τ from p_{z_0} to a point near a regular point $\tilde{p}$ of $\tilde{\Gamma}$, a small circle around $\tilde{\Gamma}$ and then back with τ^{-1}. One can construct a cylinder in $U - (\Delta \cup \tilde{\Gamma})$ whose boundaries are two circles, one being that of w_i, the other being that of w_i, the other being in $V - \Gamma$. Choose an arc $\tilde{\tau}$ in $V - \Gamma$ from p_{z_0} to that circle, and call v_i the resulting simple loop; if α_i is defined so that it follows τ and then a path along the cylinder from one circle to the other and then τ^{-1}, one realizes that $w_i = \alpha_i v_i \alpha_i^{-1}$. So b is a product of elements, each of which conjugated to an element of $\operatorname{Im} \gamma$.

COROLLARY. $\operatorname{Ker} \alpha \cap \ker \beta$ is the minimal normal subgroup N of $\ker \beta$ that contains the elements of the form $bvb^{-1}v^{-1}$ with $b \in \ker \beta$, $v \in \operatorname{Im} \gamma$.

PROOF. Let $b \in \pi_1(U - \Delta \cup \tilde{\Gamma})$, $v \in \operatorname{Im} \gamma$.

Then $bvb^{-1} = (b.\gamma\beta(b^{-1})).(\gamma\beta(b).v.\gamma\beta(b^{-1})).(\gamma\beta(b).b^{-1}) =$ $= \bar{b}.\bar{v}.\bar{b}^{-1}$ where $\bar{v} \in \operatorname{Im}\gamma$, $\bar{b} \in \ker\beta$. So if $b \in \ker\alpha$, one has from 4) and this remark that $\bar{b} = b_1 v_1 b_1^{-1} .. b_s v_s b_s^{-1}$ with $b_i \in \ker\beta$, $v_i \in \operatorname{Im}\gamma$.

Moreover $b \in \ker\beta$ implies $\beta(v_1 \dots v_s) = 1$ and hence $v_1 \dots v_s = 1$, since β is injective on $\operatorname{Im}\gamma$. Let $n_i = b_i v_i b_i^{-1} v_i^{-1} \in N$; then $b = n_1 v_1 \dots n_s v_s =$ $= v_s^{-1} \dots v_1^{-1} n_1 v_1 \dots n_s v_s$, from this and the remark that $v \in \operatorname{Im}\gamma$, $n \in N$ implies that $v^{-1} n v \in N$ one deduces $b \in N$.

REMARK. Let v be a loop in $V - \Gamma$. If we follow along v the roots of the equation defining Δ , we get an isotopy of m points in the w-plane. This is known as a braid, see E.Artin [2], which acts on the free group in m words in a natural way; this action is exactly the one studied above. Let us make some examples,

a) Δ is defined by the equation $w^2 - z^n = 0$. Then Γ is defined by $z = 0$ and $\pi_1(V - \Gamma, 0)$ has only one generator. It follows that $\pi_1(\mathbb{C}^2 - \Delta, 0)$ is presented by two generators and one set of relations induced by the generator v of $\pi_1(V - \Gamma, 0)$. For $n = 1$, when z that turns counterclockwise once around zero, $|z| = 1$, the two roots of $w^2 = z$ make half a turn. The resulting braid is pictured as .

If a and b are generators of $\pi_1(L_1 - \Delta)$ and $\tilde{a}$, $\tilde{b}$ their images under the action of v we get the following automorphism of $\pi_1(L_1 - \Delta)$: $(a,b) \longrightarrow (aba^{-1}, a)$ which is described by the following picture:

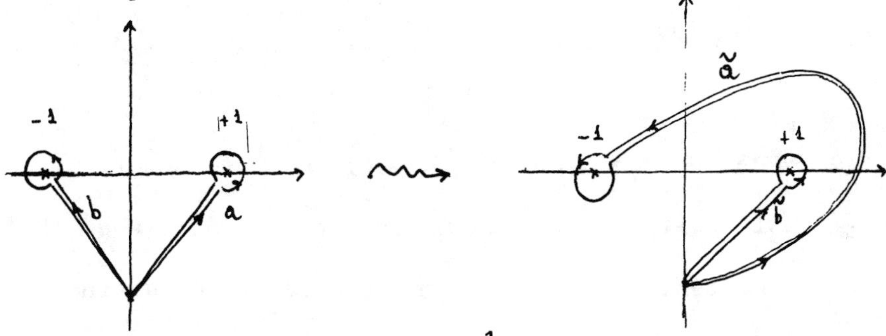

The relations we get are $a = aba^{-1}$ and $b = a$, that means $\pi_1(\mathbb{C}^2 - \Delta, 0) \simeq \{a,b \,\|\, a = b\}$ where $\{x_1,\ldots,x_r \,\|\, R_1,\ldots,R_s\}$ denotes the group generated by $x_1,\ldots,x_r$ with the relations $R_1,\ldots,R_s$.

Let $n = 2$. Then the braid associated to v is two times the braid associated to v for $n = 1$; this is pictured as it follows:

To compute the automophism of $\pi_1(L_1-\Delta,0)$ one can make the square of the automorphism corresponding to $n = 1$. We find

$$(a,b) \longrightarrow (aba^{-1}.a.ab^{-1}a^{-1}, aba^{-1}) = (abab^{-1}a^{-1}, aba^{-1})$$

from which the relations $a=abab^{-1}a^{-1}$ and $b = aba^{-1}$ of which the first ia a consequence of the second. Hence

$$\pi_1(\mathbb{C}^2 - \Delta, 0) \simeq \{a,b \parallel ab = ba\} \simeq \mathbb{Z}^2 .$$

With the same method, for $n = 3$ we find $\pi_1(\mathbb{C}^2-\Delta,0) \simeq$
$\simeq \{a,b \parallel aba = bab\}$ which is not abelian and in general:

$$\pi_1(\mathbb{C}^2 - \{w^2 = z^n\}, 0) \simeq \{a,b \parallel \underbrace{aba\ldots}_{n \text{ factors}} = \underbrace{bab\ldots\ldots}_{n \text{ factors}}\} .$$

Notice that in general $\pi_1(\mathbb{C}^h - \Delta, o)$ is presented by $m =$ (multiplicity of Δ at 0) generators. We have seen that if φ is the semiuniversal deformation of $(X, 0)$ defined by $f: (\mathbb{C}^{n+1}, 0) \to (\mathbb{C}, 0)$, then the multiplicity of Δ at 0 is equal to $\mu = \dim_{\mathbb{C}} \mathbb{C}\{x_0,\ldots,x_n\}/(\frac{\partial f}{\partial x_0},\ldots,\frac{\partial f}{\partial x_n})$

It is shown in Milnor [14] that the non singular fibre M has the homotopy type of a bouquet of μ sphere of dimension n. This can be seen in the following way: fix a generic deformation $\tilde{f} = f + \sum_0^m \alpha_i \cdot x_i$ of f (that means to consider φ over a generic line near 0 in $\mathbb{C}^h$). Then $\{|\tilde{f}|<\eta\}\cap$ $\cap\{|x|\leq\varepsilon\} = T$ for $0<\eta\ll\varepsilon\ll 1$ is contractible and $\tilde{f}: T \to \{|\lambda|<\eta\}$ has μ critical not generated points; suppose 0 is not a critical value, so that M can be identified with $\tilde{f}^{-1}(0)$. Then the Morse study of the function $|\tilde{f}|^2$ shows that T is obtained (up to homotopy equivalence) by adding to M, μ cells of dimension $n+1$.

This shows also that the μ vanishing cycles at the critical points of f, constitute a basis $e_1, \ldots, e_\mu$ of $H_n(M, \mathbb{Z})$.

At the same time, $\pi_1(\mathbb{C}^h - \Delta, 0)$ is generated by simple loops $\gamma_1, \ldots, \gamma_\mu$ corresponding to the vanishing cycles.

By the Picard-Lefshetz formula, the action of each γ_r is determined if one knows the intersection numbers $\langle e_i, e_j \rangle$. On the other hand each relation between the γ_i, can be translated (also through the Picard-Lefshetz formula) into relations between the intersection numbers.

In general one can hope to determine by this method the intersection numbers $\langle e_i, e_j \rangle$ and hence the representation $\pi_1(\mathbb{C}^h - \Delta, 0) \to \text{Aut } H_n(M, \mathbb{Z})$.

We give some examples, in the case $n-1 \equiv 0 \pmod 4$.

Then $\gamma_i(z) = z - \langle z, e_i \rangle \cdot e_i$ and

$$\gamma_i \cdot \gamma_j(z) = z - \langle z, e_i \rangle \cdot e_i - \langle z, e_j \rangle \cdot e_j + \langle z, e_j \rangle \langle e_j, e_i \rangle \cdot e_i$$

One deduces easily that $\gamma_i \gamma_j = \gamma_j \cdot \gamma_i$ iff $\langle e_i, e_j \rangle = \pm 1$

It can be shown, see [9], that for a singularity of the type $\sum_{i=0}^{n} x_i^{a_i} = 0$, this method allows us to compute the numbers $\langle e_i, e_j \rangle$; it is conjectured by E. Brieskorn that this works in general.

We have shown that Δ is irreducible and that its generic singularities are of the type $x_1 \cdot x_2 = 0$ or $x_1^2 = x_2^3$.

By translating this into properties of the presentation of $\pi_1(\mathbb{C}^h - \Delta, 0)$ one deduces (see [12]) that

a) the intersection matrix $\langle e_i, e_j \rangle$ is irreducible, that means one cannot decompose $\{1, \ldots, \mu\}$ into two sets I, J such that $\langle e_i, e_j \rangle = 0$ for $i \in I$, $j \in J$.

b) For each e_i there exists $z \in H_n(M, \mathbb{Z})$ such that $\langle e_i, z \rangle = 1$.

In general, geometric properties of Δ, reflects properties of the presentation $\pi_1(\mathbb{C}^h - \Delta, 0) \longrightarrow \mathrm{Aut}\, H_n(M, \mathbb{Z})$ or of the matrix $\langle e_i, e_j \rangle$; the same problem is found when one considers a (very ample) complete linear system on a projective variety and studies questions as "Lefschetz decomposition into invariant and vanishing cycles " or as "how many double points can have an irreducible element of the system" ; for example one expects stronger irreducibility properties (of the involved discriminant) than the simple one stated above.

[1] Andreotti, A. - Frankel, T.: "The second Lefschetz theorem on hyperplane sections" in "Global analysis", papers in honor of K. Kodaira, University of Tokyo Press, 1969.

[2] Artin, E. "Brad groups" Collected Papers.

[3] Bourbaki, N.: "Algebre Commutative, Chap. 1.", Eléments de Mathématique XXVII Hermann, Paris 1961

[4] Brieskorn, E.: "Die Monodromie der isolierten singularitaten von Hyperflachen" Manuscripta Math., v. 2 (1970), 103 - 160.

[5] Brieskorn, E. "Picard- Lefschetz theory" informal notes, not published, (1970).

[6] Frisch, J.: "Points de platitude d'un morphisme d'espaces analytiques complexes" Inv. math. 4, 118 - 136 (1967)

[7] Grauert, H.: Kerner, H.: "Deformationen von Singularitaten Komplexer Raume" Math. Ann. 153 (1964), 236 - 260

[8] Grauert, H.: "Uber die Deformation isolierter singularitaten analytischer Mengen" Inv. Math. 15 (1972) 171 - 198.

F. Lazzeri

[9] Hefez, A. - Lazzeri, F.:" **The intersection matrix of Brieskorn singularities" Inv. Math.** 25, 143-157 (1974)

[10] Kas, A.- Schlessinger, M.: "On the versal deformation of a complex space with an isolated singularity" Math. Ann. 196, 23 - 29 (1972).

[11] Kerner, H.: "Zur Theorie der Deformationen Komplexer Raume".

[12] Lazzeri, F.: "A Theorem on the monodromy of isolated singularities" astérisque 7 et 8, 1973.

[13] Lefschetz, S.: "L'Analysis situs et la géometrie algébrique", Paris, Gauthier - Villars (1924).

[14] Milnor, J.: "Singular points of complex hypersurfaces" Ann. Math. Studies, 61 Priceton.

[15] Schlessinger,M.: "Functors of Artinian Rings" Transaction AMS 130 (1968) 208 - 222.

[16] Schuster, H.W.: "Deformationen analytischer Algebren" Inv. math. 6 (1968), 262 - 274.

[17] Tjurina, G.N.: "Platte lokal semi-universelle Deformationen von isolierten Singularitaten Komplexer Raume".

Isveztia akademii naouk SSSR, Seria Mat. 33 (1969), 1026 - 1058.

[18] Von Kampen, E.R.: "On the foundamental group of an algebraic curve " Amer. J. Math., vol. 55 (1933).

[19] Whitney, H.: "Tangents to analytic varieties" Ann. of Math. 81 (1965), 496 - 549.

CENTRO INTERNAZIONALE MATEMATICO ESTIVO

(C.I.M.E.)

V. POENARU

LECTURES ON THE SINGULARITIES OF C^{∞} MAPPINGS

Corso tenuto a Bressanone dal 16 al 25 giugno 1975

V. Poénaru

SOME ELEMENTARY FACTS.

This set of lectures is an account of some aspects of Mather's theory. Some of the details will not be given in these notes, and we will sometimes refer to:

[AD] : V. Poenaru: Analyse Différentielle, Springer Lecture Notes.

We will suppose that the reader is familiar with the language of jets, germs,.....

In this first lecture some "elementary" facts will be proved. Elementary means without the differentiable preparation theorem.

1) On isolated singularities of germs of functions.

Let us consider the ring of germs of smooth functions $(R^n, 0) \to R$, $C_0^\infty(R^n)$. This is a local ring whose maximal ideal

$$\mathfrak{m} = \mathfrak{m}\, C_0^\infty(\mathbb{R}^n) = \{\text{the set of germs vanishing at } 0 \in \mathbb{R}^n\} .$$

For $f \in C_0^\infty(R^n)$ we define the jacobian ideal of f :

$J(f)$ = the ideal generated by $\frac{\partial f}{\partial x_1}, \dots, \frac{\partial f}{\partial x_n}$ in $C_0^\infty(R^n)$. Clearly, $J(f)$ does not depend on the choice of coordinates.

f will be singular ($\Leftrightarrow f$ has a singularity at the origin), if

$$J(f) \subset \mathfrak{m} .$$

Theorem 1. "Let $f \in C_0^\infty(R^n)$ be singular, and

$$g \in \mathfrak{m} J^2 \qquad (J = J(f)) .$$

Then, there exists a germ of diffeomorphism

$$(R^n, 0) \xrightarrow{\ \psi\ } (\mathbb{R}^n, 0) ,$$

such that:

$$f + g = f \circ \psi ."$$

(So $f + g$ and f differ only by a change of coordinates at the source).

Proof: Let $t \in R$. I claim that

(1) $\quad J(f+tg) = J(f)$ (equality inbetween subsets of $C_o^\infty(R^n)$).

Our hypothesis is that g is a sum of terms of the form $m_{ij} \frac{\partial f}{\partial x_i} \frac{\partial f}{\partial x_j}$, $m_{ij} \in \mathfrak{m}$; or, symbolically:

$$g = m(\partial f)^2 \quad , \text{ with } m \in \mathfrak{m} \; , \; \partial f \in J \; .$$

In the same kind of notation:

$$\partial g = (\partial f)^2 + m(\partial f)(\partial^2 f) \in \mathfrak{m} J \quad (\text{since } J \subset \mathfrak{m}) \; .$$

Hence

$$\frac{\partial(f+tg)}{\partial x_i} = \sum_j (\delta_{ij} + tm_{ij}) \frac{\partial f}{\partial x_j} \; .$$

But since $m_{ij} \in \mathfrak{m}$, the matrix $(\delta_{ij}+tm_{ij})$ is invertible, a.s.o. In fact we have found a C^∞ matrix $a_{ij}(x,t)$ such that:

$$\frac{\partial f}{\partial x_i} = \sum_j a_{ij}(x,t) \frac{\partial(f+tg)}{\partial x_j} \; .$$

Since $g \in \mathfrak{m} J^2$ we can find C^∞ functions $X_i(x,t)$, $X_i(0,t) = 0$ such that:

(2) $$g(x) = - \sum X_i(x,t) \frac{\partial(f+tg)}{\partial x_i} \; .$$

We shall think of $X_i(x,t)$ as a time-dependent vector field on $(R^n,0)$, X_t, such that

$$\begin{cases} X_t(0) = 0 \\ g = - < X_t \, , \, \partial f + t\partial g > \; . \end{cases}$$

If we <u>integrate</u> the time-dependent differential equation defined by X_t

we find a C^∞ family of (germs of) diffeomorphisms

$$(\mathbb{R}^n,0) \xrightarrow{\varphi_t} (\mathbb{R}^n, 0)$$

such that

$$\begin{cases} \varphi_o = \text{identity} \\ \frac{\partial}{\partial t}(\varphi_t) = X_t(\varphi_t) \end{cases} .$$

I claim that for $\forall t$:

(3) $\qquad f + tg = f \circ \varphi_t^{-1}$.

Proof of (3). If one starts from the general situation:

$$x \in R^n \xrightarrow{\varphi_t} R^n \xrightarrow{f_t} R \ni y$$

one has

$$T_y R \ni \frac{\partial}{\partial t}(f_t \circ \varphi_t)(x) = Tf_t(\varphi_t(x)) \circ \frac{\partial \varphi_t}{\partial t}(x) + \frac{\partial f_t}{\partial t}(\varphi_t(x)) \quad .$$

If one applies this to the special case:

$f_t = f + tg$ one has:

$$\frac{\partial}{\partial t}((f+tg)\circ \varphi_t) = (\partial f + t\partial g)(\varphi_t(x)) \circ \frac{\partial}{\partial t}\varphi_t + g \circ \varphi_t$$

(here one thinks of $\partial(f+tg)$ as a map $TR^n \to TR$ and of g as a map $\mathbb{R}^n \to TR \sim R$) $= - < X_t, \partial f + t\, \partial g > \circ\, \varphi_t + g \circ \varphi_t = 0$

(the last equality follows from (2)).

Hence $(f+tg)\circ \varphi_t$ is independent of t . Since its value for $t = 0$ is exactly f , one has q.e.d.

<u>Remarks:</u> This argument, or something similar to it is to be found in Thom's seminar at IHES and in a paper by Chenciner and Laudenbach.

Special cases:

a) "Morse functions" (non-degenerate singularities).

By definition, a germ $f \in C_0^\infty(R^n)$ which is singular at 0 has a Morse (= non-degenerate) singularity if

$$\det\left(\frac{\partial^2 f}{\partial x_i \partial x_j} (0) \right) \neq 0 \ .$$

Lemma: " f has a Morse singularity if and only if

$$\mathfrak{m} = J(f) \ ."$$

Proof: By Nakayama's lemma $\{\frac{\partial f}{\partial x_i}\}$ generates $\mathfrak{m}$ if and only if it generates $\mathfrak{m}/\mathfrak{m}^2$.

But: $\{\frac{\partial f}{\partial x_i}\}$ generates $\mathfrak{m} \Longleftrightarrow \mathfrak{m} = J(f)$ (since we deal with a singular f , and hence we have already $\mathfrak{m} \supset J(f)$).

On the other hand:

$$\frac{\partial f}{\partial x_i} - \sum_j \frac{\partial^2 f}{\partial x_i \partial x_j} (0) \, x_j \in \mathfrak{m}^2 \ .$$

Hence $\{\frac{\partial f}{\partial x_i}\}$ generates $\mathfrak{m}/\mathfrak{m}^2$ ($\Longleftrightarrow$ $\{x_j\}$ can be expressed linearily in $\frac{\partial f}{\partial x_i}$, mod $\mathfrak{m}^2$) $\Longleftrightarrow$

$$\det\left(\frac{\partial^2 f}{\partial x_i \partial x_j} (0) \right) \neq 0$$

(To see the last $\Longrightarrow$ arrow, think of $\mathfrak{m}/\mathfrak{m}^2$ as an n-dimensional R-vector space generated by the base $x_1,\ldots,x_n$) .

Corollary (Morse's lemma) = "If f is Morse and $g \in \mathfrak{m}^3$, there exists a (germ of) diffeomorphism:

$$(R^n, 0) \xrightarrow{\varphi} (R^n, 0)$$

such that

$$f + g = f \circ \varphi \ .$$

In particular, up to a change of coordinates at the source, f is completely determined by its second order Taylor expression."

b) Isolated singularities: By definition, f has an isolated singularity if $\exists k$, such that

$$\mathfrak{m}^k \subset J \ .$$

Corollary: "If f has an isolated singularity with

$$\mathfrak{m}^k \subset J \ ,$$

and if $g \in \mathfrak{m}^{2k+1}$, there exists a (germ of) diffeomorphism

$$(R^n, 0) \xrightarrow{\varphi} (R^n, 0)$$

such that

$$f + g = f \circ \varphi \ ."$$

($\mathfrak{m}^k \subset J \rightarrow \mathfrak{m}^{2k} \subset J^2$, so $g \in \mathfrak{m}^{2k+1} \rightarrow g \in \mathfrak{m} J^2$, a.s.o.)

Remark: If f is a polynomial of degree m ($\Longleftrightarrow f \in J^m$) one can always find another polynomial $F = f +$ (terms of degree $m+1$) such that F has an isolated singularity (at 0). This can be proved by complexifying and making use of the Nullstellensatz and Sard's theorem.

V. Poénaru

THE DIVISION THEOREM.

1) Statement of the main result:

THROUGHOUT THIS LECTURE $C^{\infty}(\ldots.)$ WILL DENOTE THE SPACE OF C^{∞}, COMPLEX-VALUED FUNCTIONS ON $(\ldots..)$.

We shall consider

$$(t, x) \in R \times R^n = R^{n+1}$$

$$\lambda = (\lambda_1, \ldots, \lambda_p) \in \mathbb{C}^p \quad .$$

Division theorem: "There exist two linear, continuous, smooth $(C^{0,\infty})$, mappings:

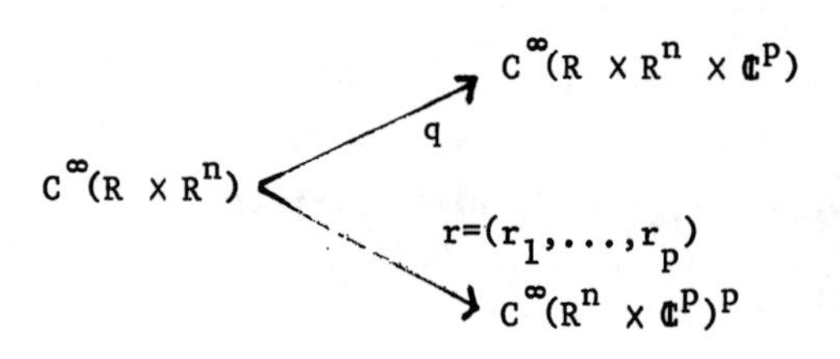

such that:

if

$$P(t, \lambda) = t^p + \lambda_1 t^{p-1} + \ldots.. + \lambda_p$$

denotes the "generic" polynomial of degree p, and

$$f(t, x) \in C^{\infty}(R \times R^n)$$

one has the identity:

$$\boxed{f(t, x) = q(t,x,\lambda)\, P(t, \lambda) + \sum_1^p r_j(x,\lambda)\, t^{p-j}}$$

where $\quad q(t,x,\lambda) = q(f)\,(t, x, \lambda)$

$$r_j(x, \lambda) = r_j(f)\,(x, \lambda) \quad ."$$

V. Poénaru

Remarks: ① Let X, Y, Z be C^∞ manifolds and T a topological space.

Any map

$$C^\infty(X) \xrightarrow{F} C^\infty(Y)$$

induces a map

$$C^{0,\infty}(T \times Z \times X) \xrightarrow{[F]} C(T \times Z \times Y)$$

where $C^{0,\infty}(T \times Z \times X) = \{$the set of continuous maps $T \times Z \times X \to \mathbb{C}$, smooth in (z, x), with all derivatives continuous in $(t, z, x)\}$ and $C(T \times Z \times Y) = \{$the set of arbitrary maps $T \times Z \times Y \to \mathbb{C}\}$.

F will be called linear, continuous, smooth if it is $\mathbb{C}$-linear and

$$\text{Image } [F] \subset C^{0,\infty}(T \times Z \times Y) .$$

② The statement above implies, trivially, an analogous statement for germs.

On the other hand, using partition of unity, one can further globalize, and replace R^n with an arbitrary smooth n-manifold M^n. The same division thm. will be true for $C^\infty(R \times M)$.

③ If one replaces

$R \times R^n \rightsquigarrow \mathbb{C} \times \mathbb{C}^n$, $C^\infty(\ldots) \rightsquigarrow C^\Omega(\ldots)$ = holomorphic functions

the division theorem remains true. It is then nothing else than the classical <u>Weierstrass preparation theorem</u>.

We give here a proof of the Weierstrass preparation theorem, which will be useful for the C^∞ case.

Consider $f(t, x)$ as a holomorphic function in t, depending on the parameter x, and write the Cauchy integral formula:

$$f(t,x) = \frac{1}{2\pi i} \int_{\partial D} \frac{f(z, x)}{z - t}\, dz \quad ,$$

where D is a disk of center 0 in the complex z-plane such that: int D contains t and all the roots of $P(z, \lambda) = 0$ for $|\lambda|$ sufficiently small.

Consider:

$$R(t, z, \lambda) = \frac{P(z, \lambda) - P(t, \lambda)}{z - t} \quad .$$

This is a holomorphic function; it is a polynomial in t (or in z), of degree $< p$.

On can express $1/z-t$ in terms of R, P :

$$\frac{1}{z-t} = \frac{P(t, \lambda)}{(z-t)\, P(z, \lambda)} + \frac{R(t, z, \lambda)}{P(z, \lambda)} \quad .$$

This really means that we have found <u>rational</u> functions (in z) $A(z,t,\lambda)$, $B_i(z,\lambda)$, such that their poles lie inside D and:

$$\frac{1}{z-t} = P(t,\lambda)\, A(z,t,\lambda) + \sum_{j=1}^{p} t^{p-j}\, B_j(z, \lambda) \quad .$$

Replacing in the Cauchy formula one gets:

$$f(t,x) = P(t,\lambda) \cdot \left[\frac{1}{2\pi i} \int_{\partial D} f(z,x)\, A(z,t,\lambda)\, dz \right] + \sum_{j=1}^{p} t^{p-j} \left[\frac{1}{2\pi i} \int_{\partial D} f(z,x)\, B_j(z,\lambda)\, dz \right] \quad .$$

(4) In the complex-analytic case one has uniqueness of q , r_j .

This is no longer true in the smooth case.

(5) In the real-analytic case only the germified version of the division theorem is true, not the global one.

2) Proof of the division theorem:

The Nirenberg-Mather extension lemma: "There exists a linear, continuous, smooth map:

$$C^{\infty}(R \times R^{n}) \xrightarrow{\quad F \quad} C^{\infty}(\mathbb{C} \times R^{n} \times \mathbb{C}^{p})$$

such that, if $f(t, x) \in C^{\infty}(R \times R^{n})$ and

$F(z, x, \lambda) = F(f)$, one has:

a) For t real :

$$F(t, x, \lambda) \equiv f(t, x) \quad .$$

b) On the set $(\mathrm{Im}\, z = 0) \cup (P(z, \lambda) = 0)$, the function $F_{\bar z}$ vanishes to infinite order."

[We use here the following notation: if $z = u + iv$, then:

$$F_{\bar z} = \frac{1}{2}\left(\frac{\partial}{\partial u} + i \frac{\partial}{\partial v} \right) F = \frac{\partial F}{\partial \bar z}$$

$$F_{z} = \frac{1}{2}\left(\frac{\partial}{\partial u} - i \frac{\partial}{\partial v} \right) F = \frac{\partial F}{\partial z} \quad .]$$

We show now how the extension lemma implies the division theorem.

One makes use of the following well-known generalization of the Cauchy formula:

$$\underbrace{F(t,x,\lambda)}_{f(t,x)} = \frac{1}{2\pi i} \int_{\partial D} \frac{F(z,x,\lambda)dz}{z - t} + \frac{1}{2\pi i} \iint_{D} F_{\bar z}(z,x,\lambda) \frac{dz \wedge d\bar z}{z - t}$$

(Here D is as before). If one uses the same device as in the complex-analytic case, one finds (at least as the result of a purely formal computation):

$$f(t,x) = P(t, \lambda)\left[\frac{1}{2\pi i}\int_{\partial D} F(z, x, \lambda)\, A(z, t, \lambda)\, dz \; + \right.$$

$$\left. + \frac{1}{2\pi i}\iint_D F_{\bar z}(z, x, \lambda)\, A(z, t, \lambda)\, dz \wedge d\bar z\right] \; +$$

$$+ \sum_{j=1}^{p} z^{p-j}\left[\frac{1}{2\pi i}\int_{\partial D} F(z, x, \lambda)\, B_j(z, \lambda)\, dz \; + \right.$$

$$\left. + \frac{1}{2\pi i}\iint_D F_{\bar z}(z, x, \lambda)\; B_j(z, \lambda)\, dz \wedge d\bar z\right].$$

There is no problem about the $\int_{\partial D}$ integrals; they are indeed C^∞ functions (in t,x,λ).

One remarks that the denominator of A is $(z-t)\, P(z,\lambda)$ and the denominator of B_j is $P(z,\lambda)$. Our assumption on F (extension lemma) is that in those points where the denominators vanish, $F_{\bar z}$ has a zero of <u>infinite</u> order. It follows that the $\iint_D$ integrals, also represent C^∞ functions (they are absolutely convergent; the integrals which one obtains by differentiating them formally, in t, x, λ , under the integral sign, are also absolutely convergent). This ends the proof.

<u>Remark:</u> The division theorem is also true for C^∞ <u>real valued</u> functions (with q, r <u>real</u>).

If f <u>is real, one</u> proceeds as before, with F replaced by $\frac{1}{2}(F(z, x, \lambda) + F(\bar z, x, \lambda))$. Then, for <u>real</u>-valued λ our integrals (divided by $2\pi i$) will be real.

3) Proof of the extension lemma:

We need a preliminary construction: Let first:

$$\delta(y, \lambda) = \inf \{ |y - \operatorname{Im} z| , P(z, \lambda) = 0 \}$$

(with $y \in R$, $\lambda \in R^p$).

THE MULTIPLIER FUNCTION: "There exists a continuous function:

$$R \times \mathbb{C}^p \times R \xrightarrow{\rho(\xi, \lambda, y)} [0, 1]$$

such that:

1°. ρ is C^∞ in λ, y and all the derivatives are continuous in (ξ, λ, y) .

2°. $\rho(\xi, \lambda, y) = 1$ in a neighbourhood of $y=0$.

3°. $|\xi y| \geq 1 \Rightarrow \rho(\xi, \lambda, y) = 0$.

4°. In a neighbourhood of $\delta(y, \lambda) = 0$, one has:

$$\rho_y(\xi, \lambda, y) = 0 .$$

5°.

$$\left| \frac{\partial}{\partial \lambda^\alpha} \frac{\partial}{\partial \bar{\lambda}^\beta} \frac{\partial}{\partial y^\gamma} \rho(\xi, \lambda, y) \right| \leq C(\lambda, y) (1 + |\xi|^K)$$

with $C(\lambda,y)$ a continuous function of (λ, y) , depending of α, β, γ , $K = K(p, \alpha, \beta, \gamma)$."

Proof:

Lemma 1: "Let

$$\Phi(x, \eta, \lambda) = \sum_{j=0}^{2p\ell-2} r_j(\eta, \lambda)x^j$$

with $r_j \in C^\infty$. Consider, for $\delta(\eta, \lambda) > 0$ the function:

$$\Psi(\eta, \lambda) = \int_{-\infty}^{\infty} \frac{\Phi(x, \eta, \lambda)\, dx}{|P(x+\eta i, \lambda)|^{2\ell}} .$$

Then:

$$\left| \frac{\partial}{\partial \lambda^{\alpha}} \frac{\partial}{\partial \bar{\lambda}^{\beta}} \frac{\partial}{\partial \eta^{\gamma}} \Psi(\eta, \lambda) \right| \leq C_1(\eta, \lambda)\, (1 + \delta(\eta, \lambda)^{-K_1})$$

with C_1 a continuous function in the variables η, λ, depending on α, β, γ ; $0 < K_1 = K_1(p, \alpha, \beta, \gamma)$."

Proof: It suffices to give the proof for the case $|\alpha| = |\beta| = \gamma = 0$ since $\frac{\partial}{\partial \lambda^{\alpha}} \frac{\partial}{\partial \bar{\lambda}^{\beta}} \frac{\partial}{\partial \eta^{\gamma}} \Psi$ is an integral of the same form as Ψ , only with a different ℓ . $|P(x+\eta i, \lambda)|^{2\ell}$ is a monic polynomial (in x or in $x + \eta i$) , of degree $2p\ell$. This implies on one hand that:

$$|P(x + \eta i, \lambda)|^{2\ell} \geq \delta(\eta, \lambda)^{2p\ell} ,$$

on the other hand that

$$|P(x + \eta i, \lambda)|^{2\ell} \geq \frac{|x|^{2p\ell}}{2}$$

when $|x|$ is larger than a certain $b = b(\eta, \lambda)$.

We decompose Ψ :

$$\Psi = \int_{-\infty}^{-b} + \int_{-b}^{b} + \int_{b}^{\infty} .$$

Let $r(\eta, \lambda) = \Sigma\, |r_j(\eta, \lambda)|$. Then:

$$\left| \int_{b}^{\infty} \right| \leq \int_{1}^{\infty} \frac{r(\eta, \lambda)\, dx}{|x|^2/2} = r(\eta, \lambda)$$

and similarly for $\int_{-\infty}^{-b}$. One also has:

$$\left| \int_{-b}^{b} \right| \leq 2\, b(\eta, \lambda)^{2p\ell-1}\, r(\eta, \lambda)\, \delta(\eta, \lambda)^{-2p\ell}$$

From here lemma 1 follows.

For $\delta(\eta, \lambda) > 0$ we consider the auxilliary function:

$$\sigma(\eta, \lambda) = \frac{1}{2\pi} \int_{-\infty}^{\infty} \left| \frac{d}{dx} \log P(x + \eta i, \lambda) \right|^2 dx \ .$$

Lemma 2: "One has:

(1) $$\frac{1}{2\,\delta(\eta, \lambda)} \leq \sigma(\eta, \lambda) \leq \frac{p^2}{2\,\delta(\eta,\lambda)}$$

(2) $$\left| \frac{\partial}{\partial \lambda^\alpha} \frac{\partial}{\partial \bar{\lambda}^\beta} \frac{\partial}{\partial \eta^\gamma} \sigma(\eta, \lambda) \right| \leq C_1(\eta,\lambda)\,(1+\delta(\eta,\lambda)^{-K_1})$$

with C_1 a continuous function in the variables η, λ, depending on α, β, γ, and

$$0 < K_1 = K_1(p, \alpha, \beta, \gamma) \ ."$$

Proof: One has: $\{z_1,\ldots,z_p\}$ = roots of $(P(z, \lambda) = 0)$, and:

$$\left| \frac{d}{dx} \log P(x+\eta i, \lambda) \right|^2 = \left| \sum_j \frac{1}{x+\eta i-z_j} \right|^2 =$$

$$= \left(\sum_j \frac{1}{x+\eta i-z_j} \right) \left(\sum_j \frac{1}{x-\eta i-\bar{z}_j} \right) .$$

By replacing x with z in the last expression, one gets a meromorphic function $Q(z)$ such that $|z|^2 Q(z)$ is bounded in the neighbourhood of ∞ .

Moreover:

$\sigma(\eta, \lambda) = i \times$ (the sum of the rezidues of Q in the upper $\frac{1}{2}$ plane).

If one develops the computations, one finds that $\sigma(\eta, \lambda)$ is a sum of p^2 quantities:

$$\beta_{jk} \frac{(\mathrm{Im}\, z_j - \eta) + (\mathrm{Im}\, z_k - \eta)}{(Rz_j - Rz_k)^2 + ((\mathrm{Im}\, z_j - \eta) + (\mathrm{Im}\, z_k - \eta))^2} \geq 0$$

with:

$$\begin{array}{lll} \beta_{jk} = 1 & \text{if} & \mathrm{Im}\, z_j, \mathrm{Im}\, z_k > \lambda \\ \beta_{jk} = -1 & \text{if} & \mathrm{Im}\, z_j, \mathrm{Im}\, z_k < \lambda \quad \text{and} \\ \beta_{jk} = 0 & & \text{otherwise.} \end{array}$$

Each of these quantities is $\leq \frac{1}{2\,\delta(\eta, \lambda)}$, and there is one of them exactly equal to $\frac{1}{2\delta(\eta, \lambda)}$ (choose $j = k = j_o$ such that

$$| \mathrm{Im}\, z_j - \lambda | = \delta(\eta, \lambda)) .$$

This proves ①. ② follows immediately from lemma 2.

Lemma 3: "Let $\rho_o : [0, \infty) \to [0, 1]$ be a C^∞ function, defined once for all, such that:

$$\begin{array}{lll} \rho_o(x) = 1 & \text{if} & x \leq 2p^3 \\ 0 \leq \rho_o(x) \leq 1 & \text{if} & 2p^3 \leq x \leq 4p^3 \\ \rho_o(x) = 0 & \text{if} & x \geq 4p^3 . \end{array}$$

Then:

$$\int_{I(\xi)} \rho_o \left(\frac{\sigma(\eta, \lambda)}{1 + |\xi|} \right) d\eta \geq \frac{1}{4(1 + |\xi|)}$$

where

$$I(\xi) = \left[\frac{1}{2(1+|\xi|)} , \frac{1}{1 + |\xi|} \right] ."$$

Proof: Let $X(\xi, \lambda) \subset R$ be the set of η's such that:

$$\eta \in I(\xi) \quad \text{and} \quad \delta(\eta, \lambda) \geq \frac{1}{4(1+|\xi|)p} .$$

I claim that:

$$\text{measure } X(\xi, \lambda) \geq \frac{1}{4(1+|\xi|)} .$$

(This inequality follows from the fact that at most $p/2$ roots of $P(z, \lambda) = 0$ have their imaginary part in $I(\xi)$; one gets $X(\xi,\lambda) \subset I(\xi)$ by excluding from $I(\xi)$ an interval of length $\frac{1}{2(1+|\xi|)p}$ around the corresponding points of $I(\xi)$).

On the other hand, from ① (lemma 2) it follows that:

$$\eta \in X(\xi, \lambda) \Rightarrow \sigma(\eta, \lambda) \leq \frac{p^2}{2\, \delta(\eta, \lambda)} \leq 2p^3(1+|\xi|)$$

$$\Rightarrow \frac{\sigma(\eta, \lambda)}{1+|\xi|} \leq 2p^3$$

$$\Rightarrow \rho_o\left(\frac{\sigma(\eta, \lambda)}{1+|\xi|}\right) \geq 1$$

$$\Rightarrow \text{q.e.d.}$$

We shall also consider a C^∞ function

$$\rho_1 : [0, \infty) \rightarrow [0, 1] ,$$

defined once for all, such that:

$$\begin{aligned} \rho_1(x) &= 0 && \text{for } x \leq \varepsilon \\ 0 \leq \rho_1(x) &\leq 1 && \text{for } \varepsilon \leq x \leq 1-\varepsilon \\ \rho_1(x) &= 1 && \text{for } x \geq 1-\varepsilon . \end{aligned}$$

We define $\rho(\xi, \lambda, y)$, for $y \geq 0$, as follows:

$$\rho(\xi, \lambda, y) = 1 \quad \text{for } y \leq \frac{1}{2(1+|\xi|)}$$

$$\rho(\xi,\lambda,y) = \rho_1 \left[\frac{\int_y^{1/1+|\xi|} \rho_o \left(\frac{\sigma(\eta,\lambda)}{1+|\xi|} \right) d\eta}{1/4(1+|\xi|)} \right]$$

for $y \in I(\xi)$

$$\rho(\xi,\lambda,y) = 0 \quad \text{for} \quad y \geq \frac{1}{1+|\xi|} .$$

One defines ρ in a similar way for $y \leq 0$. We have to check that the various properties of the MULTIPLIER FUNCTION are satisfied.

From lemma 3 and the definition of ρ_1 it follows that ρ is continuous.

1° For $y \in I(\xi)$, ρ is clearly C^∞ in the variables (λ,y) The only thing one has to check is that the partial derivatives are 0 when $y = \frac{1}{2(1+|\xi|)}$ or $y = \frac{1}{1+|\xi|}$. This follows from the fact that all the derivatives of ρ_1 are 0 for $x \leq \varepsilon$ or $x \geq 1-\varepsilon$.

2° For $y \in [-\frac{1}{2(1+|\xi|)} , \frac{1}{2(1+|\xi|)}]$, ρ is identically 1.

3° We consider $y \geq 0$ (the case $y \leq 0$ is to be treated similarly).

$$|y\,\xi| \geq 1 \Rightarrow y \geq \frac{1}{|\xi|} > \frac{1}{1+|\xi|} \Rightarrow \rho = 0$$

4° In the neighbourhood of $\delta(y,\lambda) = 0$, it follows, from lemma 2 (①) that $\sigma(y,\lambda)$ is very large ($\Rightarrow \frac{\sigma(y,\lambda)}{1+|\xi|} > 4p^3$). Now:

$$D_y \, \rho(\xi, \lambda, y) = D_{\rho_1}(\ldots) \cdot \underbrace{\left[\frac{-\rho_o \left(\frac{\sigma(y,\lambda)}{1+|\xi|} \right)}{1/4\,(1+|\xi|)} \right]}$$

this is 0 in a neighbourhood of $\delta(y,\lambda)=0$.

5° If $\gamma > 0$, one has:

$$\left| \frac{\partial}{\partial\lambda^{\alpha}} \frac{\partial}{\partial\bar{\lambda}^{\beta}} \frac{\partial}{\partial y^{\gamma}} \rho(\xi, \lambda, y) \right| \leq (\text{ "universal" constants}$$

coming from the various derivatives of $\rho_o, \rho_1) \times \frac{1}{(1+|\xi|)^N} \times$

$$\underbrace{\times\, C_1(y, \lambda)\,(1+\delta(y, \lambda)^{-K_1})}$$

this majorates the derivatives of $\sigma(y, \lambda)$.

But, if $\frac{1}{2\delta(y,\lambda)} \geq 4p^3(1+|\xi|)$

$$\Rightarrow \frac{\sigma(y,\lambda)}{1+|\xi|} > 4p^3$$

$$\Rightarrow \text{(proof of 4°)}, \quad D_y \, \rho = 0$$

$$\Rightarrow \frac{\partial}{\partial\lambda^{\alpha}} \frac{\partial}{\partial\bar{\lambda}^{\beta}} \frac{\partial}{\partial y^{\gamma}} \rho = 0 \; .$$

Hence we can replace $\delta(y, \lambda)^{-1}$ in our inequality by $|\xi|$ (and change the exponents and multiplicative constants).

If $\gamma = 0$, one has:

$$\left| \frac{\partial}{\partial\lambda^{\alpha}} \frac{\partial}{\partial\bar{\lambda}^{\beta}} \rho(\xi, \lambda, y) \right| \leq \text{(a constant coming from the derivatives}$$

of $\rho_o, \rho_1) \times$ (an upper bound of

$$\frac{\partial}{\partial\lambda^{\alpha}} \frac{\partial}{\partial\bar{\lambda}^{\beta}} \int_{y}^{1/1+|\xi|} \rho_o \left(\frac{\sigma(\eta, \lambda)}{1 + |\xi|} \right) d\eta \,) \times \frac{1}{4(1 + |\xi|)} \,.$$

Also:

$$\left| \frac{\partial}{\partial\lambda^{\alpha}} \frac{\partial}{\partial\bar{\lambda}^{\beta}} \int \right| \leq \int \left| \frac{\partial}{\partial\lambda^{\alpha}} \frac{\partial}{\partial\bar{\lambda}^{\beta}} \rho_o(\quad) \right| d\eta \leq$$

$$\leq \int C'(\eta, \lambda) \frac{1}{(1+ |\xi|)^N} (1 + \delta(\eta, \lambda)^{-K_1}) \, d\eta \,.$$

If $\dfrac{1}{2\delta(\eta, \lambda)} \geq 4p^3(1 + |\xi|)$

$$\Rightarrow \frac{\sigma(\eta, \lambda)}{1 + |\xi|} > 4p^3$$

$$\Rightarrow \rho_o\left(\frac{\sigma(\eta, \lambda)}{1 + |\xi|} \right) \equiv 0$$

$\Rightarrow$ the corresponding points (η, λ) can be excluded from the computation, and we can replace $\delta(\eta, \lambda)^{-1}$ by $|\xi|$.

This finishes the construction of the MULTIPLIER FUNCTION.

The existance of the multiplier function $\Rightarrow$ the extension lemma:

It suffices to consider the case when f has compact support; the general case follows by partition of unity. Also, the variable $x \in R^n$ will play only the role of a smooth parameter and will be omitted from our notations:

Let $\hat{f}(\xi)$ be the Fourier transform of f :

$$f(t) = \int_{-\infty}^{\infty} e^{i\xi t} \hat{f}(\xi)\, d\xi \ .$$

(The Fourier inversion formula). Remark also that $\hat{f} \in \{$ the Schwartz space $\mathcal{S}$ of rapidly decreasing functions$\}$.

Let $z \in C$ $z = x + iy$ (this is <u>not</u> the same x as before) and define F by:

$$F(z, \lambda) = \int_{-\infty}^{\infty} \rho(\xi, \lambda, y)\, e^{i\xi z}\ \hat{f}(\xi)\, d\xi \ .$$

When $|\xi\, y| \geq 1$, $\rho \equiv 0$ so this integral is uniformly absolutely convergent.

From the fact that the derivatives of ρ are bounded by a polynomial in $|\xi|$, and $|\xi\, y| \geq 1 \Rightarrow \rho \equiv 0$ it follows that

$$\left| \frac{\partial}{\partial\lambda^{\alpha}} \frac{\partial}{\partial\bar{\lambda}^{\beta}} \frac{\partial}{\partial z^{\gamma}} \frac{\partial}{\partial \bar{z}^{\delta}} \rho(\xi, \lambda, y)\, e^{i\xi z} \right|$$

is bounded (uniformly on compact sets in (y, λ)) by a polynomial in $|\xi|$. Since moreover $\hat{f} \in \mathcal{S}$, one can differentiate F without any problem under the integral sign $\Rightarrow F \in C^{\infty}$.

Since $\rho \equiv 1$ in a neighbourhood of $y = 0$:

$$F(x, \lambda) = \int_{-\infty}^{\infty} e^{i\xi x}\, \hat{f}(\xi)\, d\xi = f(x) \ .$$

$F_{\bar{z}} = 0$ of infinite order for $(y = 0) \cup (\delta(y, \lambda) = 0)$ is equivalent to:

$$\frac{\partial}{\partial\lambda^{\alpha}} \frac{\partial}{\partial\bar{\lambda}^{\beta}} \frac{\partial}{\partial z^{\gamma}} \frac{\partial}{\partial \bar{z}^{\delta}} F \Big|_{(y=0) \cup (\delta(y, \lambda) = 0)} = 0$$

for all $\delta > 0$. For $y \in (y = 0) \cup (\delta(y, \lambda) = 0)$ one has

$$\rho_y(\xi, \lambda, y) = 0 \Rightarrow \frac{\partial}{\partial \bar{z}} \rho(\xi, \lambda, y) = 0 ,$$

hence q.e.d.

<u>Final remark:</u> Since $x \in R^n$ acts only as a parameter in our proof, and everything is linear, it follows that

$$f(t, x_o) \equiv 0 \quad (\text{for } \forall\, t) \Rightarrow F(z, x_o, \lambda) \equiv 0$$

$$\Rightarrow q \text{ and } r \text{ (from the division theorem) are } \equiv 0$$

$$\text{for } x = x_o .$$

V. Poénaru

DIFFERENTIAL STABILITY.

In this section we present Mather's theorem

"infinitesimal stability ⇒ stability"

in the parametrized version due to F. Latour (C.R. 1969, pp.1331-1334). From now on, $C^{\infty}(Z)$ will denote the ring of C^{∞}, real-valued functions on the manifold Z.

We shall consider compact manifolds X, Y, and a third manifold U = the space of parameters.(In prinicple this space of parameters U should actually be R^k. Since we do not want to worry here about the standard difficulties encountered in integrating vector fields over non-compact manifolds, we shall work with compact U ; U could also be a germ of manifold. Some small extra work is required in that case; we leave it to the reader. One can also germify the whole situation: X, Y, U).

We define the U-parametrized families of smooth maps $X \to Y$:

$$C_U^{\infty}(X, Y) \subset C^{\infty}(X \times U, Y \times U)$$

as the set of $X \times U \xrightarrow{f} Y \times U$ such that the following diagramm is commutative:

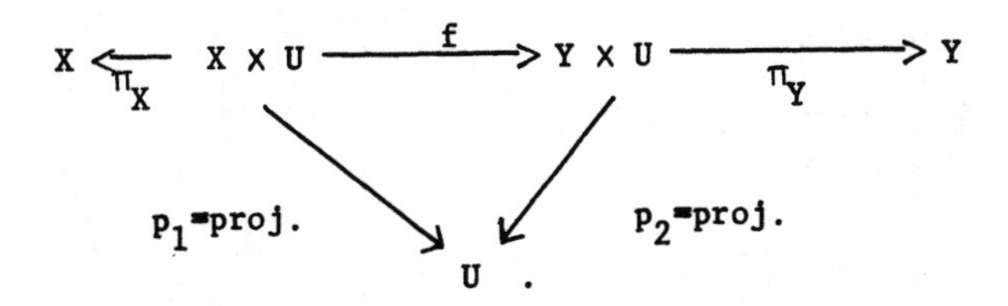

Our conventions are that:

$$C_U^{\infty}(X, Y) = C^{\infty}(U, C^{\infty}(X, Y)) \quad .$$

Let

[Diff] $(X \times U) \subset \text{Diff}\ (X \times U)$ be the set of diffeomorphisms h,

for which there exists an

$$\alpha \in \text{Diff}\ U$$

such that the following diagramm is commutative:

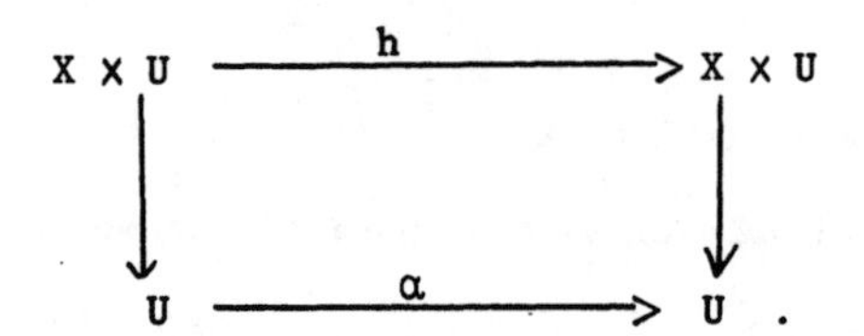

One has obvious maps:

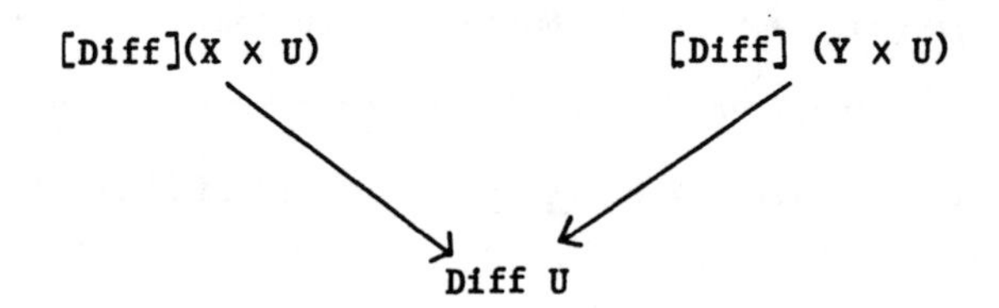

and we shall denote by G the <u>fiber</u> product:

$$G = [\text{Diff}]\ (X \times U) \underset{\text{Diff}\ U}{*} [\text{Diff}]\ (Y \times U)\ .$$

G has a natural group structure.

There is a natural group action

$$G \times C^{\infty}_{U}\ (X, Y) \xrightarrow{\Phi} C^{\infty}_{U}\ (X, Y)$$

given by

$$(h, g) \bullet f = g \circ f \circ h^{-1}\ .$$

More explicitely one has a big commutative diagramm:

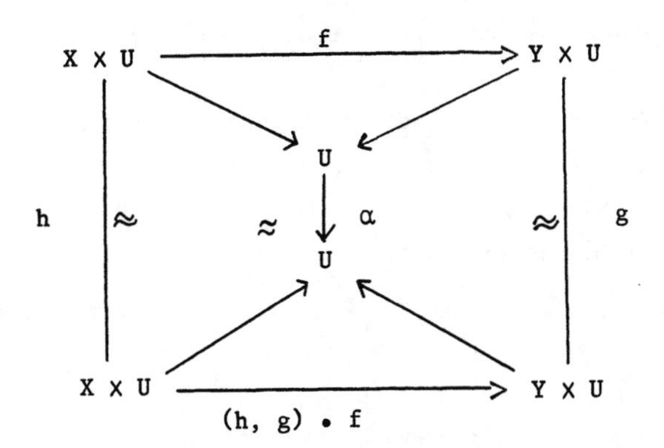

For a fixed $f \in C^{\infty}_{U}(X, Y)$, one has the orbit map:

$$\Phi_f : G \longrightarrow C^{\infty}_{U}(X, Y)$$

induced by f $(\Phi_f(\gamma) = \gamma \bullet f)$.

The image of the orbit map, is the orbit of f .

Definition: $f \in C^{\infty}_{U}(X, Y)$ is stable if there exists a neighbourhood N (in the C^{∞} topology)

$$f \in N \subset C^{\infty}_{U}(X, Y)$$

and a $C^{0,\infty}$ map Ψ such that the following diagramm is commutative:

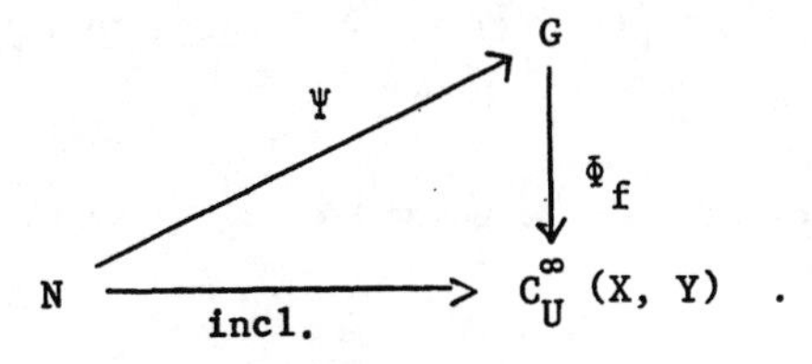

[A priori there is also another notion:

f is weakly stable if the orbit of f is a neighbourhood of f in $C^{\infty}_{U}(X, Y)$ (i.e. Ψ exists, without the $C^{0,\infty}$-property). It will turn out that the two notions are actually equivalent.]

Let $\pi_X = \pi : X \times U \to X$ be the natural projection, and

$\Gamma_U^\infty (TX) = \Gamma^\infty (\pi^* TX) = \{$the set of tangent vector fields to $X \times U$, parallel to $X \}$.

$\Gamma_U^\infty(TX)$ is a $C^\infty(X \times U)$-module, in a natural way.

We can consider the bundle $f^* \pi_Y^* TY$ over $X \times U$, and the smooth crossections

$$\eta \in \Gamma^\infty(f^* \pi^* TY) \qquad \Gamma_U^\infty (f^* TY) \qquad \text{(definition)}$$

can be identified with the smooth maps

$$X \times U \xrightarrow{\quad \eta \quad} \pi_Y^* TY$$

such that the following diagramm is commutative:

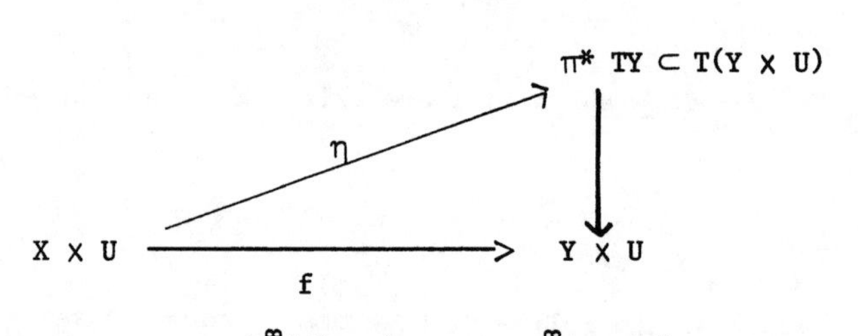

We will think of $\Gamma_U^\infty(f^* TY)$ as a $C^\infty(X \times U)$ -module. From the heuristic view-point, an element of $\Gamma_U^\infty(f^* TY)$ is an <u>infinitesimal deformation of f</u>.

One has a natural homomorphism of $C^\infty(U)$-algebras:

$$f^* : C^\infty(Y \times U) \longrightarrow C^\infty(X \times U) \quad ,$$

and, by restriction of scalars, any $C^\infty(X \times U)$-module becomes, automatically, a $C^\infty(Y \times U)$-module.

If $\xi_1 \in \Gamma_U^\infty(T X)$,

$$\beta_f(\xi_1) = Tf \circ \xi_1$$

defines a $C^\infty(X \times U)$-linear module homomorphism:

$$\Gamma_U^\infty(TX) \xrightarrow{\beta_f} \Gamma_U^\infty(f^* TY)$$

If $\xi_2 \in \Gamma_U^\infty(TY)$,

$$\alpha_f(\xi_2) = \xi_2 \circ f$$

defines a $C^\infty(Y \times R)$ -linear module homomorphism:

$$\Gamma_U^\infty(TY) \xrightarrow{\alpha_f} \Gamma_U^\infty(f^* TY) \quad .$$

One sees these maps in the diagramm below:

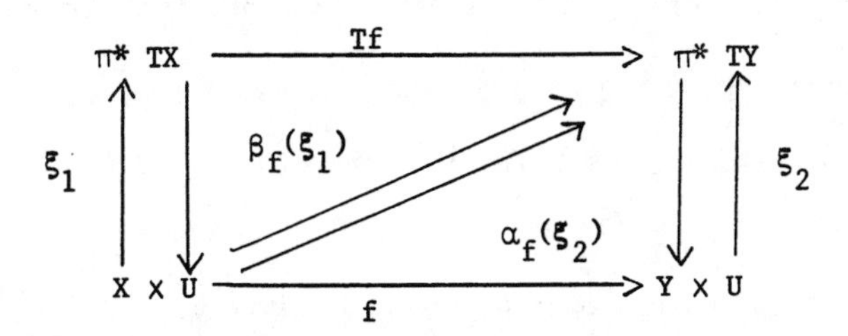

From a heuristical view-point Image β_f (resp. Image α_f) consists of those infinitesimal deformations of f coming from a U-parametrized diffeomorphism

$X \times U \to X \times U$ (resp. $Y \times U \to Y \times U$.)

If $U = pt.$ (so if one has <u>no</u> parameters), we have by now all the infinitesimal ingredients. In general there is still another item to be considered.

One also has the $C^\infty(U)$-module $\Gamma^\infty(U)$ and the map

$\tau_f : \Gamma^\infty(U) \longrightarrow \Gamma_U^\infty(f^* TY)$, defined as follows.

If $\xi_3 \in \Gamma^\infty(U)$ we have an induced field: $\pi_X^* \xi_3 \in \Gamma^\infty(T(X \times U))$, and, by definition:

$$\tau_f(\xi_3) = T\,\pi_Y \circ Tf\,(\pi_X^* \xi_3) \quad .$$

One has R-algebra homomorphisms:

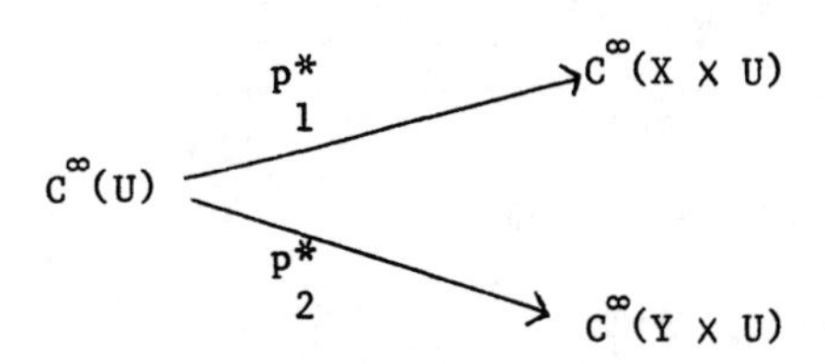

(which turn $C^\infty(X \times U)$, $C^\infty(Y \times U)$ into $C^\infty(U)$-algebras and $f^* : C^\infty(Y \times U) \to C^\infty(X \times U)$ into a $C^\infty(U)$-algebra homomorphism) and τ_f is clearly $C^\infty(U)$-linear.

From a heuristical view-point the introduction of τ_f is justified as follows:

there is a natural inclusion $\mathrm{Diff}\ U \longrightarrow G$ given by:

$$(u \to \alpha(u)) \longrightarrow ((x,u) \longrightarrow (x,\alpha(u)))\ \&\ ((y,u) \longrightarrow (y,\alpha(u))) \quad .$$

Image τ_f are exactly the infinitesimal deformations of f coming from Diff U .

Consider now 1-parameter families of diffeos h_t , α_t , g_t (equal to the identity for $t = 0$), and:

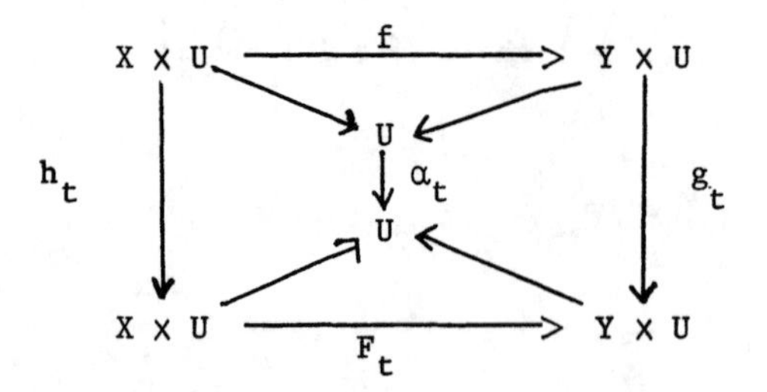

with $F_t = (h_t, g_t)\cdot f$. One can check that: $\frac{\partial F}{\partial t} \in \text{Image}\ \alpha_t + \text{Image}\ \beta_t + {}$ $+ \text{Image}\ \tau_t$. This justifies the following definition:

V. Poénaru

Definition: f is infinitesimally stable if:

$$\Gamma_U^\infty (f^* TY) = \alpha_f(\Gamma_U^\infty(TY)) + \beta_f(\Gamma_U^\infty(TX)) + \tau_f \Gamma^\infty(U) \quad .$$

STABILITY THEOREM: "If f is infinitesimally stable $\Rightarrow$ f is stable."

Remark: In all this discussion, f was a (dim U)-parameter family of smooth maps $X \to Y$. If dim U = 0 and U = pt. , $C_U^\infty(X,Y) \equiv C^\infty(X, Y)$ and one finds the usual stability concept of Thom and Mather.

1) Dynamical systems: In order to avoid complications connected with the (non) existance of global solutions for differential equations (= vector fields) we shall assume from now on that: either U is compact, or, else, U is a germ:

$$U = (U, 0) = (R^p, 0)$$

(and in this latter case with some extra work, all diffeomorphisms of U will be 0-preserving, a.s.o.)

So let X be a closed C^∞ manifold and

$$\begin{array}{c} TX \times R \\ \uparrow \; \xi \in \Gamma^\infty(\pi^* TX) \\ X \times R \end{array}$$

a (time dependent) dynamical system.

Remark also that for any

$$g \in C_R^\infty (X, Y)$$

one defines in a natural way

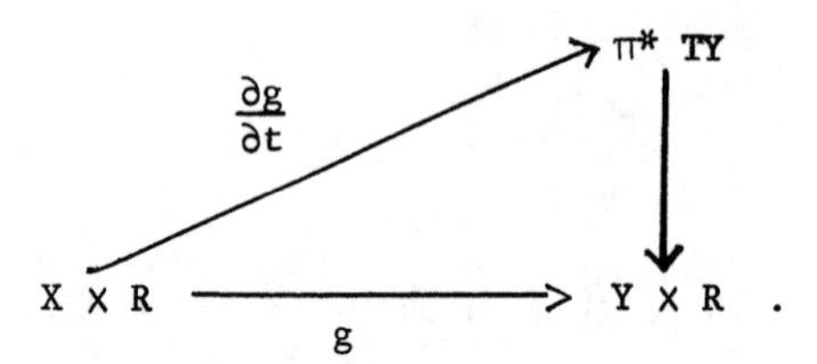

The fundamental existance theorem for differential equations tells us that to ξ there corresponds a unique

$$H \in \mathrm{Diff}\,(X \times R)$$

such that:

a) $X \times R \xrightarrow{H} X \times R$, both projecting to R (commutative triangle)

(so we think of H as an element in $C^\infty_R(X, X)$.)

b) $H(0) = id(X)$

c)

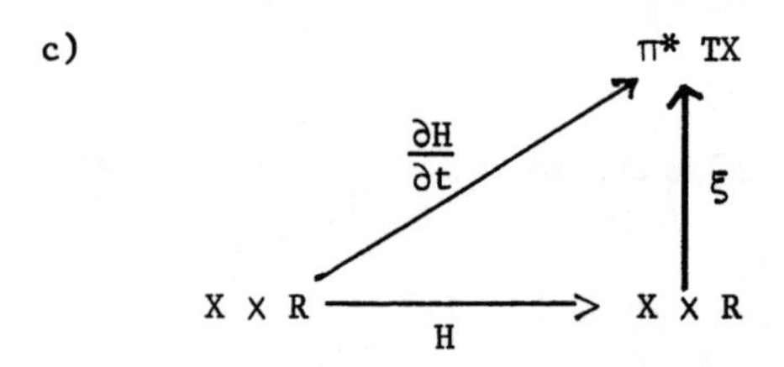

(which means: $(\frac{\partial H}{\partial t}) \circ H^{-1} = \xi$).

<u>Lemma 0.1:</u> "Let

$$F \in C^\infty_I\,(X, Y)$$

$$H \in \mathrm{Diff}\,(X \times I)$$

$$G \in \mathrm{Diff}\,(Y \times I) \quad .$$

H, G are supposed level-preserving and such that for $t = 0$ they are the identity. One considers the dynamical systems (= time dependent vector fields) corresponding to H, G :

$$\frac{\partial H}{\partial t} \circ H^{-1} = - \xi \; , \quad \frac{\partial G}{\partial t} \circ G^{-1} = \eta \, .$$

Then

$$F_t = G_t \circ F_o \circ H_t^{-1} \quad \text{for all} \quad t$$

$$\Leftrightarrow \quad \frac{\partial F_t}{\partial t} = \alpha_F(\eta) + \beta_F(\xi) \quad ."$$

This is lemma 1 (pp. 130) in [AD].

Lemma 0.2.: "Let $X_1 = X \times U$ and consider

$$\xi_t \in T(X \times U),$$

a dynamical system such that

$$\xi_t(x, u) = \underbrace{\xi_t^1(x, u)}_{\text{in } TX} + \underbrace{\xi_t^2(u)}_{\text{in } TU} \quad .$$

Let $H_t \in \mathrm{Diff}(X \times U)$, $H_t^2 \in \mathrm{Diff}\, U$ be the solutions of these dynamical systems. One has a commutative diagramm:

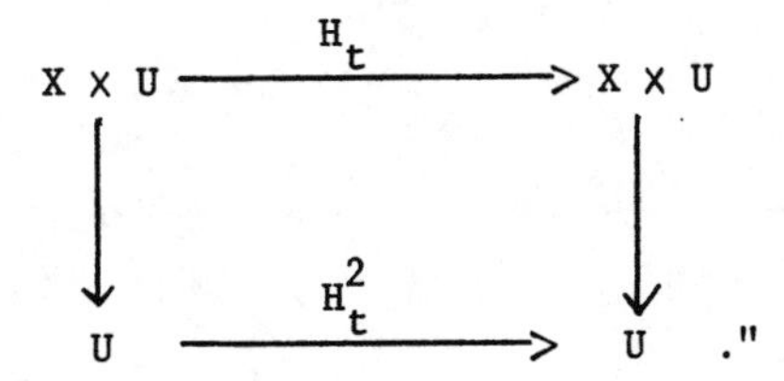

The proof is immediate.

2) The "Weierstrass property".

We consider a germ of topological space (Z_1, z_1^o) and a germ of (finite dimensional) smooth manifold (Z_2, z_2^o) . For smooth manifolds M, N we shall consider

$$C^{0,\infty}_{z_1^o, z_2^o}(Z_1 \times Z_2, C^\infty(M,N)) = C^0_{z_1^o}(Z_1, C^\infty_{z_2^o}(Z_2 \times M, N))$$

(where the notation C_z means "germs at z".)

We have a natural <u>inclusion-map</u> :

$$C^\infty(M) \hookrightarrow C^\infty_{z_1^o, z_2^o}(Z_1 \times Z_2, C^\infty(M))$$

given by the <u>constant</u> maps:

$$Z_1 \times Z_2 \longrightarrow C^\infty(M) \quad .$$

If

$$F \in C^{0,\infty}_{z_1^o, z_2^o}(Z_1 \times Z_2, C^\infty(M, N))$$

one defines an algebra-homomorphism

$$F^*_* : C^{0,\infty}_{z_1^o, z_2^o}(Z_1 \times Z_2, C^\infty(N)) \to C^{0,\infty}_{z_1^o, z_2^o}(Z_1 \times Z_2, C^\infty(M))$$

as follows: for $\Phi \in C^{0,\infty}_{z_1^o, z_2^o}(Z_1 \times Z_2, C^\infty(N))$ we set:

$$(z_1, z_2) \xrightarrow{F^*_*(\Phi)} (M \xrightarrow{F(z_1,z_2)} N \xrightarrow{\Phi(z_1,z_2)} R) \quad .$$

From now on

$$M = X \times U \quad , \quad N = Y \times U$$

and we shall fix an:

$$F \in C^{0,\infty}_{z_1^o, z_2^o}(Z_1 \times Z_2, C^\infty_U(X\ Y)) \quad .$$

<u>Theorem 1</u>: "Let us consider the following commutative square:

$$\begin{array}{ccc}
C^{0,\infty}_{z^o_1,z^o_2}(Z_1\times Z_2, C^\infty(Y\times U)) & \xrightarrow{F^*_*} & C^{0,\infty}_{z^o_1,z^o_2}(Z_1\times Z_2, C^\infty(X\times U)) \\
\downarrow {\scriptstyle ev(z^o_1,z^o_2)= ev_Y} & & \downarrow {\scriptstyle ev_X= ev(z^o_1,z^o_2)} \\
C^\infty(Y\times U) & \xrightarrow{F(z^o_1,z^o_2)^*} & C^\infty(X\times U)\ .
\end{array}$$

F^*_* is a homomorphism of $C^{0,\infty}_{z^o_1,z^o_2}(Z_1\times Z_2, C^\infty(U))$-algebras, $F(z^o_1,z^o_2)^*$ a homomorphism of $C^\infty(U)$-algebras, and the vertical maps are "algebra"-homomorphisms compatible with

$$C^{0,\infty}_{z^o_1,z^o_2}(Z_1\times Z_2, C^\infty(U)) \xrightarrow{ev_U} C^\infty(U)\ .$$

We consider also

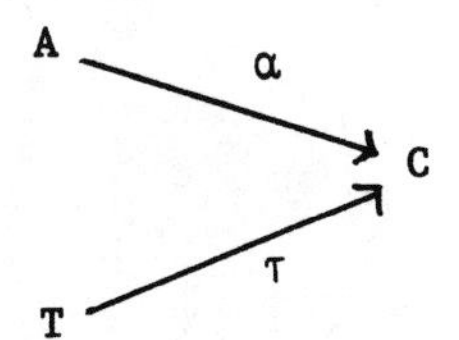

where

C is a finite $C^{0,\infty}_{z^o_1,z^o_2}(Z_1\times Z_2, C^\infty(X\times U))$-module

A is a finite $C^{0,\infty}_{z^o_1,z^o_2}(Z_1\times Z_2, C^\infty(Y\times U))$-module

T is a finite $C^{0,\infty}_{z^o_1,z^o_2}(Z_1\times Z_2, C^\infty(U))$-module,

α is $C^{0,\infty}_{z_1^o, z_2^o}(Z_1 \times Z_2, C^\infty(Y \times U))$-linear and

τ is $C^{0,\infty}_{z_1^o, z_2^o}(Z_1 \times Z_2, C^\infty(U))$-linear.

Then, the following implication holds:

$$\alpha(A) + \tau(T) + \mathrm{Ker}(ev_X)\, C = C$$

$$\Rightarrow \alpha(A) + \tau(T) = C \quad ."$$

<u>Notation:</u> Whenever there is no danger of misunderstanding we shall write z for (z_1, z_2), $C(z, x, u)$ for $C^{0,\infty}_{z_1^o, z_2^o}(Z_1 \times Z_2, C^\infty(X \times U))$, a.s.o.

<u>Lemma 1.1</u>: "$F^*_*(\mathrm{Ker}(ev_Y)) \cdot C(z, x, u) = \mathrm{Ker}(ev_X)$."

This can be proved exactly like lemma 2.1, page 135, from [AD].

<u>Definition</u>: Let us consider a ring-homomorphism $R_2 \to R_1$, R_2-modules M_2, N_2, R_1-modules M_1, N_1, and a commutative square

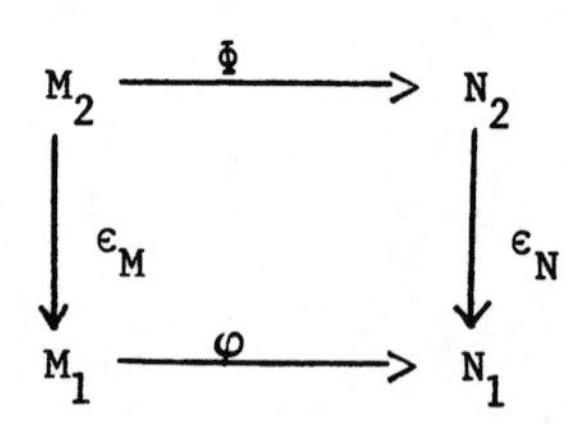

like in theorem 1. By definition, this square possesses the WEIERSTRASS PROPERTY if the following holds:

Let C be a finite N_2-module.

Then:

if $C/\Phi(\mathrm{Ker}\ \epsilon_M)$ is M_2-finite

$\Rightarrow$ C is M_2-finite. □

We also have the following lemma:

Lemma 1.2: "The commutative square from theorem 1 has the Weierstrass property."

This is lemma 2.2. (page 138) of [AD] with X, Y replaced by $X \times U$, $Y \times U$.

LEMMAS 1.1 + 1.2 $\Rightarrow$ THM. 1:

By lemma 1.1, the assumption of theorem 1 can be re-written as:

$$\alpha(A) + \tau(T) + F_*^*(\mathrm{Ker}(ev_Y)) \; C = C \quad .$$

Since A and T are finite over $C(z, y, u)$, $C(z, u)$ (which we regard as $C(z, u) \hookrightarrow C(z, y, u)$), the equality above implies that $C/F_*^*(\mathrm{Ker}(ev_Y))C$ is $C(z, y, u)$-finite.

Hence, by lemma 1.2, C is $C(z, y, u)$-finite.

From now on we forget about $C(z, x, u)$, we consider the $C(z, y, u)$-finite module

$$C_1 = C/\alpha(A)$$

and the equality:

$$(*) \qquad \tau(T) + \mathrm{Ker}(ev_Y) \; C_1 = C_1 \quad .$$

Consider $Y \times U \xrightarrow{p_2} U$, the constant map

$$Z_1 \times Z_2 \xrightarrow{p_2} C^\infty(Y \times U, U)$$

with value p_2, and the commutative square:

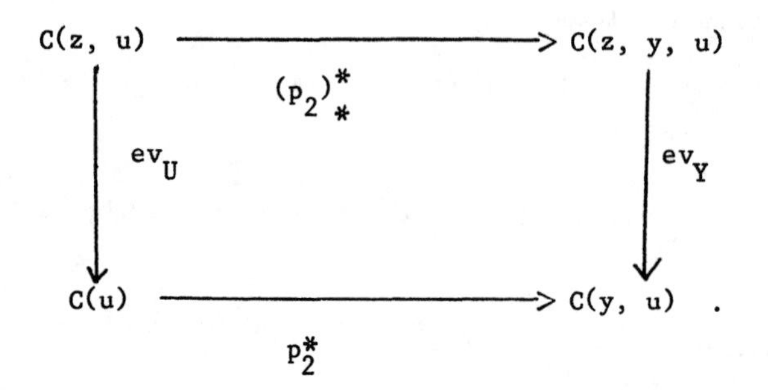

This is the square from theorem 1, with the following changes:

$$U \Longrightarrow pt$$

$$X \Longrightarrow Y \dot{\times} U$$

$$Y \Longrightarrow U$$

$$F \Longrightarrow p_2 .$$

Moreover (*) is the assumption of theorem 1, with the changes:

$$C \Longrightarrow C_1$$

$$A \Longrightarrow T$$

$$T \Longrightarrow 0 .$$

Hence, we can re-apply lemmas 1.1, 1.2 in this new context.

If we proceed as above, we deduce now that

$\underline{C_1 \text{ is } C(z, u)\text{-finite}}$,

and (*) becomes:

$$(* *) \qquad \tau(T) + \mathrm{Ker}(ev_U)\, C_1 = C_1$$

(this is an equality inbetween <u>finite</u> , C(z, u)-modules).

But

$$1 + \mathrm{Ker}(ev_U) \subset \{\text{the units of } C(z, u)\} ,$$

hence by applying Nakayama's lemma $\tau(T) = C_1$.

This is the conclusion of our theorem 1.

3) End of the proof of the stability theorem:

Let P be a compact C^∞ manifold. We consider

$$C^\infty_{U \times P}(X, Y) = C^\infty(P, C^\infty_U(X, Y))$$

and we fix some

$$F \in C^{0,\infty}_{z_1^o, z_2^o}(Z_1 \times Z_2, C^\infty_{U \times P}(X, Y)) .$$

We have an induced fiber bundle:

$$F^* TY \longrightarrow Z_1 \times Z_2 \times U \times P \times X$$

whose $C^{0,\infty}$-crossections can be identified to the maps η :

$$\begin{array}{ccc} & & Z_1 \times Z_2 \times U \times TY \\ & \overset{\eta}{\nearrow} & \downarrow \\ Z_1 \times Z_2 \times U \times P \times X & \xrightarrow[F]{} & Z_1 \times Z_2 \times U \times Y \end{array} .$$

$\Gamma^{0,\infty}(F^* TY)$ is a finite $C^{0,\infty}_{z_1^o, z_2^o}(Z_1 \times Z_2, C^\infty(U \times P \times X))$-module.

[For each $x \in U \times P \times X$ there is a neighbourhood

$$x \in V \subset U \times P \times X$$

such that

$$\Gamma^{0,\infty}(F^* TY|V) \equiv [C^{0,\infty}_{z_1^o, z_2^o}(Z_1 \times Z_2, C^\infty(V))]^N ,$$

and this implies the finiteness.]

We shall consider:

the finite $C(z, u, p, x)$-module

$$B = C^{0,\infty}_{z_1^o, z_2^o}(Z_1 \times Z_2, C^\infty(P, \Gamma^\infty_U(TX)))$$

the finite $C(z, u, p, y)$-module

$$A = C^{0,\infty}_{z_1^o, z_2^o}(Z_1 \times Z_2, C^\infty(P, \Gamma^\infty_U(T\,Y)))$$

and the finite $C(z, u, p)$-module

$$T = C^{0,\infty}_{z_1^o, z_2^o}(Z_1 \times Z_2, C^\infty(P, \Gamma^\infty(T\,U))) \quad .$$

We have maps (linear over the apropriate rings):

$$\alpha_F : A \longrightarrow \Gamma^{0,\infty}(F^* \, T\, Y)$$

$$\beta_F : B \longrightarrow \Gamma^{0,\infty}(F^* \, T\, Y)$$

$$\tau_F : T \longrightarrow \Gamma^{0,\infty}(F^* \, T\, Y)$$

defined exactly as in the context of the stability theorem (when $Z_1 = z_1^o$, $Z_2 = z_2^o$, $P = pt$). If one evaluates all these objects and maps at the origin $z^o = (z_1^o, z_2^o)$ one has exactly the data of the stability theorem, provided that $P = pt$, $F(z^o) = f$.

Lemma 2:

$$\alpha_{F(z^o)} + \beta_{F(z^o)} + \tau_{F(z^o)} \qquad \text{surjective}$$

$$\Rightarrow \alpha_F + \beta_F + \tau_F \qquad \text{surjective.}$$

Proof: We consider the commutative square:

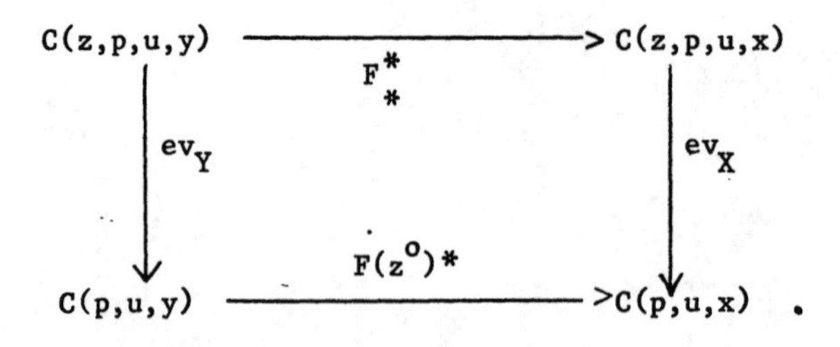

By our preceding paragraph this square has the property from Theorem 1 (replace $U \Rightarrow U \times P$).

We shall denote by $Ev(z^o)$ the natural evaluation map:

$$\Gamma^{0,\infty}(F^* TY) \xrightarrow{Ev(z^o)} \Gamma^{\infty}(F(z^o)^* TY))$$

I claim that

$$(**) \qquad \boxed{Ker\, Ev(z^o) = Ker\, ev_X . \Gamma^{0,\infty}(F^* TY)} \quad .$$

In order to prove this we can proceed exactly as in pp.158-159 of [AD], (we just stick an extra factor U , everywhere).

We also consider the finite $C(z, u, p, x)$-module

$$C = \Gamma^{0,\infty}(F^* TY)/\text{Image } \beta_F \quad \text{and}$$

the induced map

$$A \xrightarrow{\bar{\alpha}_F} C \quad .$$

The hypothesis of our lemma can be re-read as:

$$Im\, \alpha_F + Im\, \beta_F + Im\, \tau_F + Ker\, Ev(z^o) = \Gamma^{0,\infty}(F^* TY) \quad .$$

By (**) this means:

$$Im\, \alpha_F + Im\, \beta_F + Im\, \tau_F + Ker\, ev_X \cdot \Gamma^{0,\infty} = \Gamma^{0,\infty}$$

$$\Rightarrow \bar{\alpha}_F(A) + \tau_F(T) + Ker\, ev_X \cdot C = C \quad .$$

By our theorem 1 this implies

$$\bar{\alpha}_F + \tau_F \quad \text{surjective.}$$

Hence the conclusion of lemma 2 follows.

Corollary 2.1: "Let

$$f \in C^{\infty}_{U}(X, Y)$$

be infinitesimally stable.

Let Q be a compact C^{∞} manifold.

Define

$$\theta : C^{\infty}(Q, \Gamma^{\infty}_{U}(TX) \oplus \Gamma^{\infty}_{U}(TY) \oplus \Gamma^{\infty}(TU)) \to C^{\infty}(Q, \Gamma^{\infty}_{U}(f^{*} TY))$$

by:

$$Q \xrightarrow{\psi} \Gamma^{\infty}_{U}(TX) \oplus \Gamma^{\infty}_{U}(TY) \oplus \Gamma^{\infty}(TU) \xrightarrow{\beta_f + \alpha_f + \tau_f} \Gamma^{\infty}_{U}(f^{*} TY)$$

$$\theta(\psi) \ .$$

Then θ is surjective."

Proof: Consider lemma 2 with: $Z_1 = pt$, $P = pt$, (Z_2, z_2^{o}) = the germ of Q at some $z_2^{o} \in Q$, $F(z) \equiv f$.

Since f is infinitesimally stable:

$\alpha_{F(z^o)} + \beta_{F(z^o)} + \tau_{F(z^o)}$ is surjective, hence, by lemma 2:

$$C^{\infty}_{z_2^o}(Z_2, \Gamma^{\infty}_{U}(TX) \oplus \Gamma^{\infty}_{U}(TY) \oplus \Gamma^{\infty}(TU) \xrightarrow[\theta|Z_2]{} C^{\infty}_{z_2^o}(Z_2, \Gamma^{\infty}_{U}(f^{*} TY))$$

is surjective.

Our result follows now from a C^{∞} partition of 1 on Q .

Corollary 2.2: "Let

$$f \in C^{\infty}_{U}(X, Y)$$

be infinitesimally stable.

We consider a germ of topological space (Z, z^{o}) and some

$$F \in C^{0}_{z^o}(Z, C^{\infty}(I, C^{\infty}_{U}(X, Y)))$$

such that $F(z^{o})$ is the constant path

$$I \to f \ .$$

Then:

$$C^{o}_{z^o}(Z, C^{\infty}(I, \Gamma^{\infty}_{U}(TX)) \oplus C^{o}_{z^o}(Z, C^{\infty}(I, \Gamma^{\infty}_{U}(TY))) \oplus$$

$$\oplus C^{o}_{z^o}(Z, C^{\infty}(I, \Gamma^{\infty}(TU))) \xrightarrow[\beta_F+\alpha_F+\tau_F]{} \Gamma^{0,\infty}(F^* TY)$$

is surjective."

Proof: Consider the preceding corollary, with $Q = I$. The map θ is then nothing else than

$$\alpha_{F(z^o)} + \beta_{F(z^o)} + \tau_{F(zo)} .$$

Lemma 2 and corollary 2.1 imply that

$$\alpha_F + \beta_F + \tau_F \text{ is surjective.}$$

Lemma 3: "Let

$$f \in C^{\infty}_{U}(X, Y) .$$

There exists a neighbourhood

$$f \in N \subset C^{\infty}_{U}(X, Y)$$

and a continuous map

$$F : N \longrightarrow C^{\infty}(I, C^{\infty}_{U}(X, Y))$$

such that:

If $g \in N$ ', $F(g)$ is a smooth path starting with f and ending with g ."

This is an easy exercise (we suppose here that U is a manifold. If U is just a germ one needs a slightly more careful rewording.)

Suppose now that f is infinitesimally stable, and N, F are like in lemma 3. Let (Z, z^o) be the germ of N at $f \equiv z^o$, and $t \in I$.

One thinks of

$$F \in C^0_{z^0}(Z, C^\infty(I, C^\infty_U(X, Y)))$$

and defines $\frac{\partial F}{dt} \in \Gamma^{0,\infty}(F^* TY)$. This is, of course an arrow:

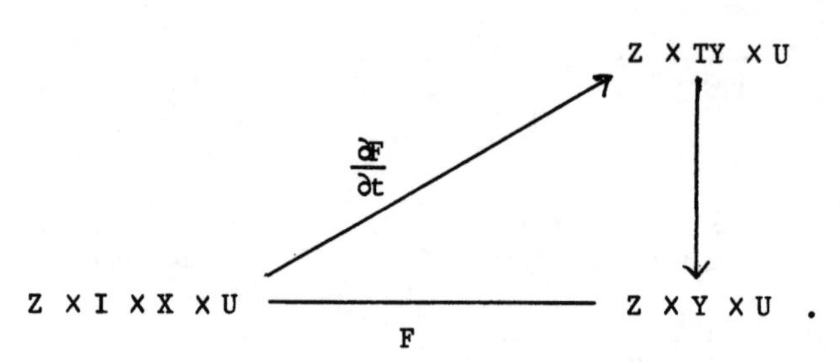

Since f is infinitesimally stable, corollary 2.2 implies the existance of vector fields:

$$\xi \in C^0_{z^0}(Z, C^\infty(I, \Gamma^\infty_U(TX)))$$

$$\eta \in C^0_{z^0}(Z, C^\infty(I, \Gamma^\infty_U(TY)))$$

$$\theta \in C^0_{z^0}(Z, C^\infty(I, \Gamma^\infty(U)))$$

such that

$$\alpha_F(\eta) + \beta_F(\xi) + \tau_F(\theta) = \frac{\partial F}{\partial t} .$$

In particular, for $g \in N$ close to f , and for the path

$$F(g) \in C^\infty(I, C^\infty_U(X, Y))$$

one has:

$$\alpha_{F(g)}(\eta) + \beta_{F(g)}(\xi) + \tau_{F(g)}(\theta) = \frac{\partial F(g)}{\partial t} .$$

(One thinks here of η , ξ, θ as elements of $C^\infty(I, \Gamma^\infty_U(TY)),\ldots$)

More explicitely, one has

$$\eta_t(y, u) \in T_y X$$

$$\xi_t(x, u) \in T_x X$$

$$\theta_t(x, u) \equiv \theta_t(u) \in T_u U$$

and

$$\underbrace{\eta_t \circ F_t}_{\eta_t(F_t(x,u),u)} + TF_t(\xi_t(x,u)) + \text{Proj}(TY \times TU \to TY) TF_t\ \theta_t(x,u) = \frac{\partial F_t}{\partial t}$$

We think here of

$$\eta_t \circ F_t \in TY \times 0 \subset TY \times TU$$

$$TF_t(\xi_t) \in TY \times 0 \subset TY \times TU$$

$$TF_t\ \theta_t \in TY \times TU\ .$$

But F_t has the form:

$$\begin{cases} y = F_t(x,\ u) \\ u = u \end{cases}$$

hence its jacobian matrix is:

$$\begin{pmatrix} \frac{\partial F}{\partial x} & \frac{\partial F}{\partial u} \\ 0 & \text{Id} \end{pmatrix} .$$

This means that:

$$TF_t\ \theta_t = \underbrace{\text{Proj}(TY \times TU \to TY) TF_t\ \theta_t}_{\text{in } TY} + \underbrace{\theta_t \circ F_t}_{\text{in } TU} .$$

We consider now the vector-fields:

$$\xi' = \xi - \theta \quad \text{in} \quad TX \times TU$$

$$\eta' = \eta + \theta \quad \text{in} \quad TY \times TU$$

and we get:

$$\frac{\partial F_t}{\partial t} = \eta'_t \circ F_t + TF_t \circ \xi'_t\ .$$

One can conclude the proof of the structural stability theorem, by appealing to lemmas 0.1 and 0.2.

V. Poénaru

(UNI)-VERSAL UNFOLDING.

There are several versions of the theory of (uni)-versal unfolding; for example:

Turina: On flat semi-universal deformations...(Russian) Izv. Akad. Nauk URSS 1969 pp. 1026-1058

Thom: Modèles mathématiques de la morphogénèse, IHES 1971.

Lassalle: (to appear)

Sergeraert: Un theorem funct. impl..... Amer. Ec.N.Sup. 1972

Mather: (to appear)

The version we give here is that of V.M. Zakalyoukin. (Funk. Anal. t.7 1973, pp. 28-31).

We consider:

$C_o^\infty(R^n)$ = the local ring of germs of C^∞ functions around $0 \in R^n$.

$C_o^\infty(R^n, R^m) = C_o^\infty(R^n)^m$ = the $C_o^\infty(R^n)$ module of germs of C^∞ maps $(R^n, 0) \to R^m$.

We shall consider some fixed

$$f \in C_o^\infty(R^n, R^m) .$$

A(k-dimensional)-unfolding of f , is by definition, an

$$F \in C_o^\infty(R^n \times R^k , R^m) ,$$

such that $F(x, 0) \equiv f(x)$.

DEFINITION 1: The unfolding F is called VERSAL , if for any other unfolding of f :

$$G \in C_o^\infty(R^n \times R^\ell, R^m)$$

the following holds: there exist C^∞ (germs of) maps

$$(R^n \times R^\ell, 0) \xrightarrow{\quad S \quad} (R^n \times R^k, 0)$$
$$(R^\ell, 0) \xrightarrow{\quad \sigma \quad} (R^k, 0)$$

such that: 1) The following diagramm is commutative:

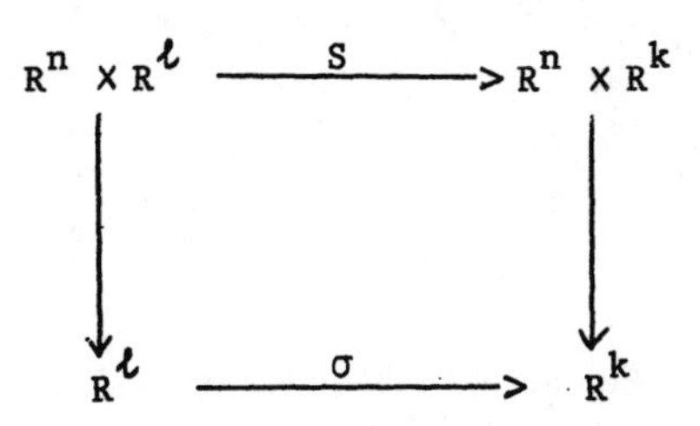

2) $S | R^n \times 0 \equiv \mathrm{id}\,(R^n)$.

3) G is <u>induced</u> from F via S :

$G = F \circ S$.

More explicitely, the following diagramm is commutative:

$$R^n \times R^\ell \xrightarrow{\quad S \quad} R^n \times R^k \xrightarrow{\quad F \quad} R^m .$$

(with $G : R^n \times R^\ell \to R^m$ the composite.)

In the $C_o^\infty(R^n)$-module $C_o^\infty(R^n, R^m)$ we will consider the <u>jacobian sub-module of f</u> $J(f) \subset C_o^\infty(R^n, R^m)$, which is, by definition:

$$J(f) = \{ \frac{\partial f}{\partial x_1}, \ldots, \frac{\partial f}{\partial x_n} \} \, C_o^\infty(R^n)$$

(i.e. the submodule generated by the $\partial f / \partial x_i \in C_o^\infty(R^n, R^m)$) . It is clear $J(f)$ is independent of the coordinate system (x_i) . If $m = 1$, $C_o^\infty(R^n, R) \equiv C_o^\infty(R^n)$ and $J(f)$ is the <u>jacobian ideal of f</u> .

Let $u_1, \ldots, u_k$ be local coordinates around 0 in R^k .

DEFINITION 2: The unfolding F of f is called INFINITESIMALLY VERSAL, if the following holds:

The R-vector space

$$C_o^\infty(R^n, R^m)/J(f)$$

is generated by the images of:

$$\left.\frac{\partial F}{\partial u_1}\right|_{u=0}, \ldots\ldots, \left.\frac{\partial F}{\partial u_k}\right|_{u=0} .$$

THEOREM: "If F is an infinitesimally versal unfolding of f, then F is a versal unfolding of f."

Example: Let $f \in C_o^\infty(R^n)$ basing an isolated singularity at the origin. This means that

$$\exists N \text{ , such that } \mathfrak{m}^N \subset J(f)$$

or, equivalently, that

$$\dim_R C_o^\infty(R^n)/J(f) < \infty .$$

Let $\varphi_1(x), \ldots, \varphi_k(x) \in C_o^\infty(R^n)$ be generators of $C_o^\infty(R^n)/J(f)$. Then the unfolding

$$F(x, u) = f(x) + \sum_1^k u_i \varphi_i(x)$$

is versal.

Proof of the theorem: Let F be infinitesimally versal, as above. Consider some other k-dimensional unfolding of f:

$$G \in C_o^\infty(R^n \times R^k, R^m) .$$

We shall assume that G is such that:

$$\left.\frac{\partial(G-F)}{\partial u_j}\right|_{u=0} \in J(f) .$$

More explicitely, this means:

$$\left.\frac{\partial(G-F)}{\partial u_j}\right|_{u=0} = \sum_s \left.\frac{\partial F}{\partial x_s}\right|_{u=0} \cdot R_{sj}(x) = \sum_s \frac{\partial f}{\partial x_s} R_{sj}(x)$$

with $R_{sj}(x) \in C_o^\infty(R^n)$.

[In heuristical terms: infinitesimally, G and F differ by a diffeo of the source.]

We will consider

$$G_t(x, u) = F(x, u) + t(G(x,u) - F(x, u))$$

$(t \in [0, 1])$. Note that G_t is a germ of map $R^n \times R^k \times I \to R^m$, along $0 \times I$:

$$G_t \in C^\infty_{o \times I}(R^n \times R^k \times I, R^m) \text{ , and that: } G_o \equiv F \text{ , } G_1 \equiv G \text{ .}$$

Lemma 1: "There exist germs of C^∞ functions:

$$X_i \in C^\infty_{o \times I}(R^n \times R^k \times I) \quad (i=1,\dots,n)$$

$$E_j \in C^\infty_{o \times I}(R^k \times I) \quad (j=1,\dots,k)$$

such that:

a) $X_i(x, 0, t) = 0$, $E_j(0, t) = 0$.

b) One has:

$$\boxed{F(x,u) - G(x,u) = \sum_{i=1}^n X_i(x, u, t) \frac{\partial G_t}{\partial x_i} + \sum_{j=1}^k E_j(u, t) \frac{\partial G_t}{\partial u_j}} \text{ ."}$$

Proof: Let

$$\Phi(x, u) \in C_o^\infty(R^n \times R^k)$$

There are operators:

$$A_j : C_o^\infty(R^n \times R^k) \to C_o^\infty(R^n \times R^k)$$

(j=1,...,h) such that:

$$\Phi(x, u) = \Phi(x, 0) + \sum_{j=1}^{k} u_j A_j(\Phi)(x, u) \quad .$$

[One can take for A_j the following:

$$A_j \Phi(x, u) = \int_0^1 \frac{\partial}{\partial u_j} \Phi(x, tu)\, dt \quad .]$$

We shall introduce auxiliary parameters:

$$\delta \in R^k \quad , \lambda \in R^k$$

and consider the following elements in the module

$$C_0^\infty(R^n \times R^{3k}, R^m) \quad :$$

$$M_i(x,u,\delta,\lambda) = \frac{\partial F}{\partial x_i}(x,u) + \sum_{r=1}^{k} \delta_r A_r\left(\frac{\partial G}{\partial x_i} - \frac{\partial F}{\partial x_i}\right)(x, u)$$

$$N_j(x,u,\delta,\lambda) = \frac{\partial F}{\partial u_j}(x,u) + \sum_r \delta_r A_r\left(\frac{\partial G}{\partial u_j} - \frac{\partial F}{\partial u_j}\right)(x, u) -$$

$$- \sum_{s,r} \delta_r R_{sj}(x) A_r\left(\frac{\partial F}{\partial x_s}\right)(x,u) - \sum_{s,r} \lambda_r R_{sj}(x) A_r\left(\frac{\partial G}{\partial x_s} - \frac{\partial F}{\partial x_s}\right)(x,u) \quad .$$

One has: $(t \in [0, 1])$

$$M_i(x,u,tu,t^2u) = \frac{\partial F}{\partial x_i} + \sum tu_j A_j\left(\frac{\partial G}{\partial x_i} - \frac{\partial F}{\partial x_i}\right) = \frac{\partial G_t}{\partial x_i} .$$

(since $G(x, 0) - F(x, 0) \equiv 0$), and in particular:

$$M_i(x, 0, 0, 0) = \left.\frac{\partial F}{\partial x_i}\right|_{u=0} \quad .$$

Also:

$$N_j(x,u,tu,t^2u) = \left[\frac{\partial F}{\partial u_j} + \sum_r tu_r A_r\left(\frac{\partial G}{\partial u_j} - \frac{\partial F}{\partial u_j} \right) + \right.$$

$$\left. + t \left.\frac{\partial(G-F)}{\partial u_j}\right|_{u=0} \right] + \underbrace{\left[-t \left.\frac{\partial(G-F)}{\partial u_j}\right|_{u=0}\right.}_{t \sum_s R_{sj}(x) \left.\frac{\partial F}{\partial x_s}\right|_{u=0}} - \sum_{s,r} tu_r R_{sj}(x) A_r \left(\frac{\partial F}{\partial x_s}\right) -$$

$$\left. - \sum_{s,r} t^2 u_r R_{sj}(x) A_r \left(\frac{\partial G}{\partial x_s} - \frac{\partial F}{\partial x_s}\right)\right] = \frac{\partial G_t}{\partial u_j} - t \sum_s R_{sj} \frac{\partial G_t}{\partial x_s}$$ and

$$N_j(x, 0, 0, 0) = \left.\frac{\partial F}{\partial u_j}\right|_{u=0} .$$

$M_i(x, u, \delta, \lambda)$, $N_j(x, u, \delta, \lambda)$ were germs at $0 = (0,0,0,0)$. But $M_i(x, u, tu, t^2u)$, $N_i(x, u, tu, t^2u)$ make sense for all $t \in [0, 1]$ and are, hence, germs along $0 \times I$.

$(M_i, N_j(x, u, tu, t^2u) \in C^\infty_{o \times I}(R^n \times R^k, R^m))$.

[For the structure of the proof below, it is essential that we have constructed $M_i(x, u, \delta, \lambda)$, N_j , such that $M_i(x,u,tu,t^2u)$, $M_i(x, 0,0,0)$ is as above, and $N_j(x, u, tu, t^2u) = \frac{\partial G_t}{\partial u_j} + \sum_i B_{ij}(x, x) \frac{\partial G_t}{\partial x_i}$, $N_j(x, 0, 0, 0) =$ as above.]

We consider

$$C_1 = C^\infty_o(R^n \times R^{3k}, R^m)/ \{M_i\} C^\infty_o(R^n \times R^k, R^m) .$$

This is a finite $C^\infty_o(R^n \times R^{3k})$-module.

We consider the projection

$$R^n \times R^{3k} \to R^{3k}$$

and the natural R-algebra homomorphism:

$$C^\infty_o(R^n \times R^{3k}) \longleftarrow C^\infty_o(R^{3k}) .$$

This makes out of C_1 a $C_o^\infty(R^{3k})$-module.

I claim that

$$\dim_R C_1 / \; C_o^\infty(R^{3k}) \, . \, C_1 < \infty$$

and that, moreover $N_1, \ldots, N_k$ generate $C_1 / \; C_o^\infty(R^{3k}). \, C_1$ as an R-vector space.

[Indeed, if

$$\Phi(x, u, \delta, \lambda) \in C_o^\infty(R^n \times R^{3k}, R^m) \quad ,$$

one has:

$$\Phi(x,u,\delta,\lambda) - \Phi(x,0,0,0) \in \mathfrak{m}\, C_o^\infty(R^{3k}) \cdot C_o^\infty(R^n \times R^{3k}, R^m) \quad .$$

Since F is infinitesimally versal:

$$\Phi(x, 0,0,0) = \Sigma \frac{\partial f}{\partial x_i} \underbrace{h_i(x)}_{\text{in } C_o^\infty(R^n)} + \Sigma \frac{\partial F}{\partial u_j}\Big|_{u=0} \underbrace{c_j}_{\text{in } R} =$$

$$= \Sigma \frac{\partial F}{\partial x_i}\Big|_{u=0} h_i(x) + \Sigma \frac{\partial F}{\partial u_j}\Big|_{u=0} c_j =$$

$$= \Sigma\, M_i(x, 0, 0, 0)\; h_i(x) + \Sigma\, N_j(x, 0, 0, 0)\; c_j \quad .$$

Hence:

$$\Phi(x,0,0,0) - \Sigma\, M_i(x,u,\delta,\lambda) h_i(x) - \Sigma\, N_j(x,u,\delta,\lambda) c_j \in \mathfrak{m}\, C_o^\infty(R^{3k}).C_o^\infty(R^n \times R^{3k}, R^m) \;,$$

a.s.o.]

Using the <u>differentiable preparation theorem</u> we deduce that

C_1 is a $C_o^\infty(R^{3k})$-finite module,

and generated (as a $C_o^\infty(R^{3k})$-module) by the N_j's.

Hence, every $\Phi(x, u, \delta, \lambda)$ can be expressed as:

$$\Phi = \Sigma\, M_i \; h_i(x, u, \delta, \lambda) + \Sigma\, N_j \; c_j(u, \delta, \lambda) \quad .$$

We can do this for

$$\Phi = F(x, u) - G(x, u) \quad ,$$

and replace δ by tu, λ by t^2u .

This means:

$$F\text{-}G = \Sigma \frac{\partial G_t}{\partial x_i} \tilde{h}_i(x, u, tu, t^2u) +$$

$$+ \Sigma \left(\frac{\partial G_t}{\partial u_j} - t \sum_s \frac{\partial G_t}{\partial x_s} R_{sj}\right) \tilde{c}_j (u, tu, t^2u) =$$

$$= \Sigma \frac{\partial G_t}{\partial x_i} \underbrace{(\tilde{h}_i - t \sum_r R_{ir} \tilde{c}_r)}_{X_i} + \Sigma \frac{\partial G_t}{\partial u_j} \underbrace{\tilde{c}_j}_{E_j} \quad .$$

This is equation b) (from the statement of our lemma). In order to have a) one applies the following changes to the former procedure:

Since $F(x, 0) - G(x, 0) \equiv 0$, one can write:

$$F(x,u) - G(x, u) = \sum_j u_j \Phi_j(x,u) \quad , \text{ with } \Phi_j = A_j(F\text{-}G) \quad .$$

One applies our preceding argument to each $\Phi_j(x, u)$; so:

$$\Phi_j(x,u) = \sum_i \frac{\partial G_t}{\partial x_i} X_i^j(x,u,t) + \sum_k \frac{\partial G_t}{\partial u_k} E_k^j(u, t) \quad ,$$

and finally:

$$X_i(x, u, t) = \sum_j u_j X_i^j(x, u, t)$$

$$E_k(u, t) = \sum_j u_j E_k^j(u, t) \quad .$$

With these new definitions, both b) <u>and</u> a) are clearly fulfilled.

<u>Lemma 2:</u> "Let G be as above. There exist germs of diffeomorphisms, preserving 0 (and such that

$$(S|u = 0) \equiv \text{identity in } x .)$$

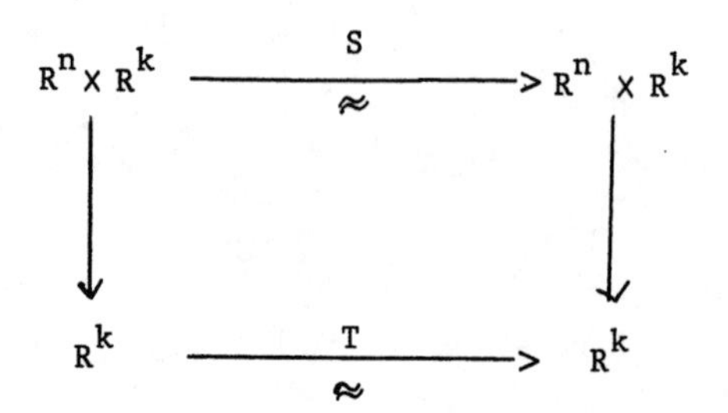

such that $F = G \circ S$."

Proof: Consider lemma 1 and $X_t = (X_j(t))$, thought of as a time dependent vertical vector field on $R^n \times R^k$, $E_t = (E_i(t))$, thought of as a time dependent vector field on R^k .

One has $E_t(0) = 0$, $X_t(x,0) \equiv 0$.
(E_t) can be lifted to $R^n \times R^k$ and there is a one-parameter family of diffeomorphisms of $R^n \times R^k$ (keeping the origin fixed) S_t , such that:

$$\frac{\partial S_t}{\partial t} = (X_t, E_t) \circ S_t \quad . \qquad (S_o = id)$$

S_t covers the diffeomorphisms T_t which solve the equation:

$$\frac{\partial T_t}{\partial t} = E_t \circ T_t \quad . \qquad (T_o = id)$$

Moreover

$$S_t | (u=0) \equiv \text{identity in } x \quad .$$

I claim that

$$\boxed{F = G_t \circ S_t} \quad .$$

This is obviously true for $t = 0$. So we have to prove that:

$$\frac{\partial}{\partial t} (G_t \circ S_t) = 0 \quad .$$

But, by the "chain rule" for differentiating a composition of maps:

$$\frac{\partial}{\partial t}(G_t \circ S_t) = \frac{\partial G_t}{\partial t} \circ S_t + TG_t(S_t) \circ \frac{\partial S_t}{\partial t} =$$

$$= \left(\frac{\partial G_t}{\partial t} + \langle TG_t , (X_t, E_t) \rangle \right) \circ S_t$$

Hence we must show that:

$$(*) \qquad \frac{\partial G_t}{\partial t} + \langle TG_t , (X_t, E_t) \rangle = 0 \quad .$$

But $\frac{\partial G_t}{\partial t} = G - F$ and:

$$\langle TG_t , (X_t, E_t) \rangle = \Sigma X_i \frac{\partial G_t}{\partial x_i} + \Sigma E_j \frac{\partial G_t}{\partial u_j} \quad ,$$

hence (*) is nothing but equation b) from lemma 1. So, lemma 2 is proved.

<u>End of the proof of the versal unfolding theorem:</u>

Let

$$H \in C_o^\infty(R^n \times R^\ell , R^m)$$

be some unfolding of f .

We consider the unfolding $\tilde{H} \in C_o^\infty(R^n \times R^{k+\ell}, R^m)$, of f :defined by:

$$\tilde{H}(x, u, \lambda) = H(x, \lambda) + \sum_1^k u_j \left. \frac{\partial F}{\partial u_j} \right|_{u=0} \quad .$$

Since F is infinitesimally versal,

$$\left. \frac{\partial \tilde{H}}{\partial \lambda_s} \right|_{\lambda=0} = \left. \frac{\partial H}{\partial \lambda_s} \right|_{\lambda=0} = h_1^s + h_2^s \quad ,$$

with $h_1^s = \Sigma c_j \left. \frac{\partial F}{\partial u_j} \right|_{u=0}$

$$h_2^s = \Sigma a_i(x) \left. \frac{\partial F}{\partial x_i} \right|_{u=0} \in J(f) \quad .$$

We define the unfolding of f :

$$F_1 \in C_o^\infty(R^n \times R^{k+\ell}, R^m)$$

by:

$$F_1(x, u, \lambda) = f(x) + \Sigma u_j \frac{\partial F}{\partial u_j}_{u=0} + \Sigma \lambda_s h_1^s \quad .$$

I claim that <u>F_1 is infinitesimally versal.</u>

This follows immediately from the equality:

$$\frac{\partial F_1}{\partial u_j}\Big|_{u,\lambda=0} = \frac{\partial F}{\partial u_j}\Big|_{u=0} \quad .$$

By re-expressing the h_1^s in terms of $\frac{\partial F}{\partial u_j}\Big|_{u=0}$, one obtains a linear map

$$R^\ell \xrightarrow{\ L\ } R^k \quad , \qquad (L = (L_1,\ldots,L_k))$$

such that:

$$F_1(x, u, \lambda) = F(x, 0) + \Sigma \frac{\partial F}{\partial u_j}\Big|_{u=0} (u_j + L_j(\lambda)) \quad .$$

Now: H can be induced from $\tilde{H}$ by the embedding

$$R^n \times R^\ell \equiv R^n \times 0 \times R^\ell \subset R^n \times R^k \times R^\ell \quad .$$

$\tilde{H}$ and F_1 are like G and F in lemma 2.

[This can be seen as follows:

From: $\frac{\partial \tilde{H}}{\partial \lambda_s}\Big|_{(u=\lambda=0)} = \frac{\partial H}{\partial \lambda_s}\Big|_{\lambda=0} = h_1^s + h_2^s$, it follows that:

$$\frac{\partial}{\partial \lambda_s}(\tilde{H} - F_1) = h_2^s \in J(f) \quad .$$

On the other hand:

$$\frac{\partial}{\partial u_j}(\tilde{H} - F_1) = \frac{\partial F}{\partial u_j}\Big|_{u=0} - \frac{\partial F}{\partial u_j}\Big|_{u=0} = 0 \in J(f) \quad .]$$

So $\widetilde{H}$ can be induced from F_1 (via a "global", parametrized diffeomorphism).

If we consider

$$F_2(x, u) = F(x, 0) + \Sigma u_j \left.\frac{\partial F}{\partial u_j}\right|_{u=0} ,$$

then F_1 can be induced from F_2 by

$$(\lambda, u) \rightarrow (u + L(\lambda)) .$$

Finally F_2 and F are like G and F in lemma 2, so F_2 can be induced from F.

So our initial H can be induced from F, and our theorem is proved.

Heuristic interpretation: If we consider the group G of germs of diffeomorphism $(R^n, 0) \rightarrow (R^n, 0)$ acting on $C_o^\infty(R^n, R^m)$, we can view

$$C_o^\infty(R^n, R^m) / J(f)$$

as a vector space in $T_f C_o^\infty(R^n, R^m)$, transversal to the G-orbit of f. Our infinitesimal condition can be viewed as meaning transversality to the G-orbit of f.

V. Poénaru

Appendix: TRANSVERSALITY THEORY.

We include a brief exposition of the Thom-Abraham transversality theorem without Banach-manifolds. As before in these notes we consider only the compact case (C^∞-topology). With some extra work things can be made to work for the non-compact case (C^∞-fine, or Whitney-topology)

If X, Y, Z are smooth manifolds, then, by definition:

$$C^\infty(Z, C^\infty(X, Y)) \equiv C^\infty(Z \times X, Y) \quad .$$

A subset $M \subset C^\infty(X, Y)$ is called a WEAK MANIFOLD if it satisfies the following conditions:

A-0) Any

$$f \in C^\infty([0, \infty), M)$$

can be extended to an

$$F \in C^\infty_k((-\infty, \infty), M) \quad .$$

A-1) Let c^k_f denote the constant map:

$$I^k \longrightarrow f \in C^\infty(X, Y) \quad .$$

If $f \in C^\infty(X, Y)$ and U is some neighbourhood:

$$c^1_f \in U \subset C^\infty(I \times X, Y) = C^\infty(I, C^\infty(X, Y)) \quad ,$$

there exists a neighbourhood

$$f \in V \subset C^\infty(X, Y) \quad ,$$

such that any $g \in V \cap M$ can be connected to f by a (smooth) arc in $U \cap C^\infty(I, M)$.

A-2) Let $f \in M$. We give some neighbourhood

$$f \in V \subset M \quad .$$

There exists a neighbourhood

$$c_f^1 \in U \subset C^\infty(I, M)$$

such that if $\psi_1, \ldots\ldots, \psi_k \in U$, $\psi_i(0) = f$, we can find a

$$\Psi \in C^\infty(I^k, V)$$

with the properties:

a) $\Psi(0) = f$

b) $\Psi(0,\ldots,0, t_i, 0,\ldots,0) = \psi_i(t_i)$.

A-3) Let $f \in M$, $\Psi \in C^\infty(I^k, M)$ with the property $\Psi(0) = f$.

For any neighbourhood

$$\Psi \in W \subset C^\infty(I^k, M) ,$$

there exists a neighbourhood:

$$c_f^1 \in U \subset C^\infty(I, M)$$

such that for any $\psi \in U$ with $\psi(0) = f$, there is a

$$\Psi_1 \in C^\infty(I \times I^k, M)$$

such that:

a) $\Psi_1(0, u) = \Psi$

b) $\Psi_1(t, 0) = \psi$

c) for any fixed $t_o \in I$, $\Psi_1(t_o,\ldots) \in W$.

Trivially, any smooth submanifold of $C^\infty(X, Y)$ (in particular $C^\infty(X, Y)$ itself) is a weak manifold.

For any $f \in C^\infty(X, Y)$ we consider the R-vector space

$$T_f\, C^\infty(X,Y) = \Gamma^\infty(f^* TY) \qquad \text{(definition).}$$

If $f \in M$,

$$T_f\, M \subset T_f\, C^\infty(X, Y)$$

is, by definition the subset of all

$$\frac{\partial F}{\partial t}\Big|_{t=0} \in T_f\, C^{\infty}(X, Y) \quad ,$$

where $F \in C^{\infty}(I, M)$, $F(0) = f$. Our axioms are such that $T_f M$ is a (sub)-vector space.

For any $x \in X$, $f \in M$ there is a natural linear map:

$$ev(x, f) : T_f\, M \to T_{f(x)} Y \quad .$$

If $Z \subset Y$ is some given submanifold, we say that M is TRANSVERSAL to Z if for every $x \in X$, $f \in M$ such that $f(x) \in Z$, one has:

$$T_{f(x)} Z + Tf(T_x\, X) + ev(x, f)\, T_f\, M = T_{f(x)}\, Y \quad .$$

If $M = \{f\}$ this reduces to the usual notion of f TRANSVERSAL to Z

<u>THEOREM:</u> "If M is transverse to Z , the set of $g \in M$ which are transversal to Z is an open dense set (in M)."

(This is the compact case. In general, the set of transversal elements will be a countable intersection of open dense set. M will be assumed to have the Baire property.)

Proof: Transversality is clearly an "open condition", and everything follows if one shows that:

(II) "If $x \in X$, $f \in M$, there are neighbourhoods: $x \in V_x \subset X$, $f \in V_f \subset M$, such that the set of $g \in V_f$ with the property that $g|V_x$ is transverse to Z , is dense in V_f ."

Let $f \in M$, $x \in X$ as above. The fact that M is transversal to Z and axiom A-2) imply the existence of some (Ψ, k) :

$\Psi \in C^\infty(I^k, M)$

such that: i) $\Psi(0) = f$

ii) the map

$$I^k \times X \xrightarrow{\Psi} Z$$

is transversal to Z at $(0, x)$.

Without loss of generality (take a smaller k-cube) $\Psi | I^k \times V_x$ is transversal to Z , for some compact neighbourhood $x \in V_x \subset X$. Let W be some neighbourhood

$\Psi \in W \subset C^\infty(I^k, M)$

such that for any $\Phi \in W$, one has:

$\Phi | I^k \times V_x$ is transversal to Z .

We apply axiom A-3) to this W , and find, hence, a $C_f^1 \in U \subset C^\infty(I, M)$ and

$\Phi_1 \in C^\infty(I^{k+1}, M)$.

Mark that $\Phi_1 | I^{k+1} \times V_x$ is still transversal to Z . By axiom A-1), there is a neighbourhood:

$$f \in V_f \subset M$$

such that any $g \in V_f$ can be joined to f by an arc in U . Hence, for any $g \in V_f$ we can construct Φ_1 as before, with

$$\Phi_1(t_o, x) \equiv g(x)$$

for some value $t_o \in I^{k+1}$. We shall show that the set

$$Reg = \{t \in I^{k+1} , \Phi_1(t) | V_x \text{ transversal to } Z\}$$

is dense in I^{k+1} . (This clearly implies proposition (II)). Since

$\Phi_1 | I^{k+1} \times V_x$ is

transverse to Z ,

$$(\Phi_1 | I^{k+1} \times V_x)^{-1} Z \subset I^{k+1} \times V_x$$

is a smooth submanifold

$$S \subset I^{k+1} \times V_x \ .$$

We can consider the commutative diagramme:

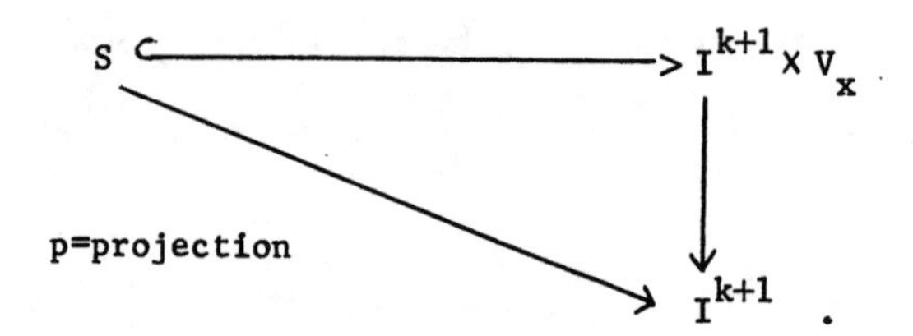

I claim that if $u \in I^{k+1}$, $y \in V_x$ and $(u, y) \in S$ (hence $\Phi_1(u, y) \in Z$) then

(*) u is a regular value of $p \Longrightarrow \Phi_1$ is transversal to Z , at (u,y).

(This remark plus Sard's theorem $\Rightarrow$ Reg is dense in I^{k+1}) .

So, let $u \in I^{k+1}$ be a regular value of p and $(u,y) \in S$. This means that $T_u I^{k+1}$, as a subspace of $TI^{k+1} \oplus TV_x$ is contained in

$$T_{(u,y)} S + T_y V_x \subset T_u I^{k+1} \oplus T_y V_x \ .$$

Since Φ_1 is transversal to Z . The following composition of mappings is onto:

$$\underbrace{T_u I^{k+1} \oplus T_y V_x}_{= T_{(u,y)}(I^{k+1} \times V_x)} \xrightarrow{T\Phi_1} T_{\Phi_1(u,y)} Y \longrightarrow T_{\Phi_1(u,y)} Y / T_{\Phi_1(u,y)} Z \ .$$

Since, by definition, the image of TS by this composition of maps is 0,

V. Poénaru

$$\text{Image } T\, I^{k+1} \hookrightarrow \text{Image } TV_x$$

$$\Longrightarrow T_y V_x \longrightarrow T_{\Phi_1(u,y)} Y / T_{\Phi_1(u,y)} Z \longrightarrow 0$$

$$\Longrightarrow \Phi_1(u) | V_x \text{ is transversal to } Z .$$

q.e.d.

Some references for Transversality theory:

[1] R. Thom: Un lemme sur lès applications différentiables. Bul. Soc. Mat. Mex. (1956) pp. 59-71

[2] C. Morlet: Le théorème de transversalité. Sem. H. Cartan 1961-1962.

[3] R. Abraham: Transversal mappings and flows. Benjamin 1967.

[4] J. Mather: Stability of C^∞ mappings V , Transversality Adv. in Mathematics (1970) vol 4 pp.301-336.

[5] V. Poénaru: Sur la structure des sphères d'homotopie lisses en dim. 3, I (preprint, Orsay 1971) [The present text is borrowed from [5] and this in turn is taken from some mimeographed notes, by the author, from Harvard 1964].

CENTRO INTERNAZIONALE MATEMATICO ESTIVO

(C. I. M. E.)

A. TOGNOLI

ABOUT THE SET OF NON COHERENCE OF A REAL ANALYTIC VARIETY

PATHOLOGY AND IMBEDDING PROBLEMS FOR REAL ANALYTIC SPACES

Corso tenuto a Bressanone dal 16 al 25 giugno 1975

About the set of non coherence of a real analytic variety

by A. Tognoli* (Pisa)

Introduction. Let (X, O_X) be a real analytic variety and $B(X)$ the set of non coherent points of X.

W.Fensh has proved that $B(X)$ has codimension at least two, in [2] is shown that $B(X)$, in general, is not an analytic subset of X.

M.Galbiati (see [3]) has proved that there exists a stratification of X by semianalytic sets such that $B(X)$ is a union of strata and hence $B(X)$ is semianalytic in X.

In this lecture we give a different proof of the result of M. Galbiati and W. Fensh.

* Lavoro eseguito nell'ambito del G.N.A.S.A.G.A. del C.N.R. .

A. Tognoli

§ 1. Preliminary remarks.

In the following we shall consider analytic subsets of $\mathbb{R}^n$ and we shall use the terminology of the previous lecture about the non coherent sets of the I and the II kind. More shortly we shall call an analytic subset of the I (or the II) kind a non coherent analytic subset of the I (or the II) kind.

Let X be a real analytic subset of $\mathbb{R}^n$ and suppose that X is coherent or of the I kind. Let $\{\tilde{X}_\lambda\}_{\lambda\in\Lambda}$ be the family of complex analytic subsets of some neighbourhood of X in $\mathbb{C}^n$ such that $\tilde{X}_\lambda \cap \mathbb{R}^n = X$. Let $\tilde{X}^g_{x_0}$ be the germ of the complex analytic subset defined by $\bigcap_{\lambda\in\Lambda} \tilde{X}_{\lambda,x_0}$ (*) (the intersection $\bigcap_{\lambda\in\Lambda} \tilde{X}_{\lambda,x_0}$ is equivalent to a finite intersection and $\tilde{X}^g_{x_0}$ is a well defined germ that contains, canonically, the complexification $\tilde{X}_{x_0}$ of X_{x_0}).

Lemma 1. Let X be an analytic subset of $\mathbb{R}^n$ and suppose X coherent or of the I kind.
There exists a neighbourhood $\tilde{U}$ of X in $\mathbb{C}^n$ and an analytic subset $\tilde{X}^g$ of $\tilde{U}$ such that:

i) $\tilde{X}^g_x = (\tilde{X}^g)_x$ for any $x\in X$ and $\tilde{X}^g \cap \mathbb{R}^n = X$

Moreover X is coherent iff $\tilde{X}^g_x \simeq \tilde{X}_x$ for any $x \in X$.

(*) If V is an analytic set .... by $V_x, x\in V$, we denote the germ of V in the point x and by $\tilde{V}_x$ the complexification of V_x.

Proof. We must verify that the germs $\tilde{X}^g_x$ glue together into an analytic space; it is then enough to prove that any $\tilde{X}^g_x$ induces $\tilde{X}^g_y$ if y is sufficiently near to x.

Let $\tilde{V}^x$ be a realisation of $\tilde{X}^g_x$, it is clear that (if y is near to x) $\tilde{V}^x_y \supset \tilde{X}^g_y$.

It is known (see [6]) that x has, in $\tilde{V}^x$, a neighbourhood U_x such that any analytic component W of $\tilde{V}^x$ that intersects U_x contains x. Let us suppose that for any neighbourhood $D \subset U_x$ there exists y such that $\tilde{X}^g_y \neq \tilde{V}^x_y$, then $\tilde{V}^x_y$ has an irreducible component ${}^i\tilde{V}^x_y \not\subset \tilde{X}^g_y$ but any realization of ${}^i\tilde{V}^x_y$ in D contains x. Then if $\tilde{X}'$ is a complex space such that $\tilde{X}' \supset X$, $\tilde{X}'$ induces $\tilde{X}^g_y$, we have $\tilde{X}'_x \subsetneq \tilde{X}^g_x$ and this is impossible. So we have proved $\tilde{V}^x_y = \tilde{X}^g_y$ and the lemma is true.

Remark 1. A different proof of lemma 1 is the following, let $\{\mathcal{J}_\lambda\}_{\lambda\in\Lambda}$ be the family of the coherent ideal sheaves of $O_{\mathbb{R}^n}$ such that: support $O_{\mathbb{R}^n}/\mathcal{J}_\lambda = X$. let $\mathcal{J}$ be the subsheaf generated by $\bigcup_\lambda \mathcal{J}_\lambda$. it is known (see [7]) that $\mathcal{J}$ is coherent (and $\sqrt{\mathcal{J}} = \mathcal{J}$), then the space $\tilde{X}^g$ constructed in lemma 1 is a complexification associated to the analytic space $(X, O_{\mathbb{R}^n}/\mathcal{J})$.

<u>Definition 1</u>. Let X be an analytic subset of $\mathbb{R}^n$ and suppose that X be coherent, or not coherent of the I kind. We shall call a completely reduced complexification of X any space $\tilde{X}^g$ defined in lemma 1.

<u>Remark 2</u>. Given $X \subset \mathbb{R}^n$ the completely reduced complexification is uniquely determined as a germ near X (so in particular the germs $\tilde{X}^g_x$, $x \in X$ are determined).
The completely reduced complexification depends on the embeddings of X in $\mathbb{R}^n$ (see [4]) and we have $\dim_{\mathbb{R}} X = \dim_{\mathbb{C}} \tilde{X}^g$ (see [5]).
Let X be an analytic subset of $\mathbb{R}^n$, $(\tilde{X}^g, \mathcal{O}_{\tilde{X}^g})$ a completely reduced complexification and $(X, \mathcal{O}_{X^g})$ the real analytic space defined by the ideal $\mathcal{J} = (\bigcup_\lambda \mathcal{J}_\lambda)$ (see remark 1).
For any $x \in X$ we have: (1) $\mathcal{O}_{\tilde{X}^g,x} \simeq \mathcal{O}_{X^g,x} \otimes \mathbb{C}$, and let: $\mathcal{J}_x \otimes \mathbb{C} = \bigcap_{i=1}^{g} \tilde{I}^i_x$ be the decomposition of $\mathcal{J}_x \otimes \mathbb{C}$ into the prime factors (we recall that, by the construction: $\tilde{\mathcal{J}}_x = \mathcal{J}_x \otimes \mathbb{C} = \sqrt{\mathcal{J}_x \otimes \mathbb{C}}$).

Let ${}^i\tilde{X}^g_x$ be the irreducible germ associated to $\tilde{I}^i_x$.

<u>Definition 2</u>. A germ ${}^i\tilde{X}^g_x$ (or the prime factor $\tilde{I}^i_x$) of $\tilde{X}^g_x$ is called of the first type iff $\dim_{\mathbb{C}} {}^i\tilde{X}^g_x = \dim_{\mathbb{R}}({}^i\tilde{X}^g_x \cap \mathbb{R}^n)$.
We shall call ${}^i\tilde{X}^g_x$ (or $\tilde{I}_x$) <u>of the second type</u> iff it is not of the first type.

<u>Definition 3</u>. Let X be an analytic subset of $\mathbb{R}^n$, we shall say that $x \in X$ <u>satisfies the condition</u> ${}_pB(X)$ (or $x \in {}_pB(X)$) iff the exists at least one irreducible component ${}^iX_x$ of X_x such that: $\dim {}^iX_x = p$ and for any realization $\tilde{V}^i$ of ${}^i\tilde{X}_x$ there exists $y \in \tilde{V}^i \cap \mathbb{R}^n$ and an irreducible component $\tilde{V}_y^{i,k}$ of $\tilde{V}_y^i$ such that $\dim(\tilde{V}_y^{i,k} \cap \mathbb{R}^n) < p$.

<u>Lemma 2</u>. <u>Let</u> X <u>be a pure dimensional analytic subset of</u> $\mathbb{R}^n$ <u>non coherent of the</u> I <u>kind</u>.
<u>For any</u> $x \in X$ <u>we have</u>: X_x <u>has one irreducible component of the I type non coherent iff</u> x <u>satisfies the condition</u> ${}_pB(X), p = \dim X$.

<u>Proof</u>. We remember the following fact; let V_x be a germ of an analytic set, then V_x is coherent iff the complexification $\tilde{V}_x$ of V_x induces the complexification of V_y for any y sufficiently near to x (see [8]).
Let now $x \in {}_pB(X)$ then the complexification ${}^i\tilde{X}_x$ of ${}^iX_x$ doesn't induces the complexification of ${}^i\tilde{X}_x \cap \mathbb{R}^n$ in any neighbourhood of x and hence ${}^iX_x$ is not coherent. We know that $\dim {}^iX_x = \dim X$ hence ${}^iX_x$ is of the first type.
Let now $x \notin {}_pB(X)$ then any component of the I type ${}^iX_x^g$ of X_x^g is coherent because, by dimensional reasons, the complexification ${}^i\tilde{X}_x = {}^i\tilde{X}_x^g$ of ${}^iX_x = {}^i\tilde{X}_x \cap \mathbb{R}^n$ induces the complexifi

cation in a neighbourhood.

Remark 3. We have now a method to detect the set $B(X)$ of non coherent points of an analytic set.

It is known that $x \in B(X)$ iff at least one irreducible component X_x^i of X_x is not coherent. Then we have $x \in B(X)$ iff there exists X_x^i such that $q = \dim X_x^i$ and for any realisation $\tilde{V}^i$ of $\tilde{X}_x^i$ there exists $y \in V^i = \tilde{V}^i \cap \mathbb{R}^n$ and $\tilde{V}_y^{i,k}$ component of $\tilde{V}_y^i$ such that $\dim(\tilde{V}_y^{i,k} \cap \mathbb{R}^n) < q$.(*)

In other words X_x^i is not coherent iff x satisfies the condition $_qB(X)$, $q = \dim X_x^i$. If X is pure dimensional and $p = \dim X$ it is possible to describe the set $_pB(X)$ using $\tilde{X}^g$ (see lemma 2).

We give now an example of an analytic subset X of $\mathbb{R}^4$ such that: X is not coherent of the I kind and the set $B(X)$ is not an analytic subset of X .

Let $V = \{(x,y,z,w) \in \mathbb{R}^4 \mid x^3 - x^2wz - wy^2 = 0\}$

it is not difficult to verify that (see [2]):

(*) We remark that X_x^i is a component of X_x and not the real part of some ${}^{i}\tilde{X}_x^g$. In fact ${}^{i}\tilde{X}_x^g \cap \mathbb{R}^n$ may be embedded in some other X_x^j and also if ${}^{i}\tilde{X}_x^g \cap \mathbb{R}^n$ satisfies the condition of definition 3 for some $q < p$ the point x may be coherent.

1) V is irreducible

2) $B(V) = \{(x,y,z,w) | x=0, y=0, w=0, z \geq 0\} \cup \{(x,y,z,w) | x=0, y=0, z=0\}$.

§ 2. The main theorem.

Let (X, O_X) be an analytic variety we shall note:

$X_{q-reg} = \{x \in X / x$ has a neighbourhood isomorphic to $\mathbb{R}^q\}$. If X is of the I kind and $p = \dim X$ we have, by theorem B, $x \in X_{p-reg}$ iff $\tilde{X}_x^g$ is smooth.

Definition 4. For any analytic variety (X, O_X) of dimension p we shall write:

$X(1) = X - \bar{X}_{p-reg}$ (handle)

$X(2) = \bar{X}_{p-reg} \cap \overline{X(1)}$ (handle connection)

$X(3) = \bar{X}_{p-reg} - (X_{p-reg} \cup \overline{X(1)})$ (embedded handle).

The terminology is clarified by the following example (Whitney's umbrella):

$$X = \{(x,y,z) \in \mathbb{R}^3 / x^2 - zy^2 = 0\}$$

we have:

$$X(1) = \{(0,0,z) | z<0\}, \quad X(2) = (0,0,0), \; X(3) = \{(0,0,z) \; z>0\}.$$

In general it results:

$x \in X(1)$ iff $\dim X_x < p$

$x \in X(2)$ iff $\dim X_x = p$ but there exists an x_n such that, $x_n \longrightarrow x$ and $\dim X_{x_n} < p$.

$x \in X(3)$ iff $\dim X_x = p$, $\dim X_y = p$ for any y near enough to x and $x \notin X_{p\text{-reg}}$.

If $X \subset \mathbb{R}^n$ is of the I kind we note X_s the singular locus of $X^g (X_s = (\text{sing } \tilde{X}^g) \cap \mathbb{R}^n)$.

It is easy to verify that:

1) $X_s = X(1) \cup X(2) \cup X(3)$ and $X(i) \cap X(k) = \emptyset$ if $i \neq k$

2) $X(1)$ and $X(3)$ are open in X_s and $X(2)$ is closed.

Lemma 3. *Let* $X \subset \mathbb{R}^n$ *be an analytic subset of* $\mathbb{R}^n$ *then* $X(1), X(2), X(3)$ *are semianalytic in* X.

Proof. The question is local so we may suppose X of the I kind. It is known (see [9]) that $X_{p\text{-reg}}$, and hence $\bar{X}_{p\text{-reg}}$, is semianalytic in X. It is now clear that $X(1), X(2), X(3)$ are semianalytic in X.

Lemma 4. *Let* X *be an analytic subset of the I kind of* $\mathbb{R}^n$ *then:* codimension $X(2) \geq 2$.

Proof. There exist a stratification $X = \bigcup_n A_n$ of X such that any $X(i)$ is the union of strata (see [9]).
It is clear that $\dim X(1) < \dim X$; $X(2)$ is a union of strata A'_n that are contained in the closure of the strata of $X(1)$. Then $\dim X(1) > \dim X(2)$ and the lemma is proved.

We generalise definition 4 with the following:

Definition 5. For any analytic variety we note:

$_pX(1) = X - \bigcup_{q\geq p} \bar{X}_{q-reg}$ (handle of dimension less than p)

$_pX(2) = \overline{_pX(1)} \cap (\overline{\bigcup_{q\geq p} X_{q-reg}})$ (connection of handle $_pX(1)$)

$_pX(3) = X_s - (_pX(1) \cup {_pX(2)})$ (embedded handle).

Lemma 5. Let (X, O_X) be an analytic normal variety and $B(X)$ the locus of non-coherent points.

There results:

1) $B(X) = \bigcup_p {_pX(2)}$ then $B(X)$ is semianalytic in X.

2) $\dim X - \dim B(X) \geq 3$.

Proof. The problem is local then we may assume that X be an irreducible analytic set of the I kind of $\mathbb{R}^n$.

It is known (see [5]) that $x \in B(X)$ iff there exists $x_n \longrightarrow x$ such that $\dim X_{x_n} < \dim X_x$. So $B(X) = \bigcup_p {_pX(2)}$ is proved.

By the same arguments used in lemma 3 it follows that $_pX(2)$ are semianalytic, hence, by local finiteness of $\{_pX(2)\}_{p\in\mathbb{N}}$ we deduce that $B(X)$ is semianalytic.

It is known (see [5]) that $\dim X - \dim X_1 \geq 2$ (X is normal) hence by the same argument used in lemma 4 we deduce $\dim X - \dim B(X) \geq 3$.

We can now prove the main theorem:

Theorem 1. Let (X,O_X) be a real analytic variety and $B(X)$ the set of non coherent points.

We have:

1) $B(X)$ is a closed semianalytic set of X.

2) $\dim X - \dim B(X) \geq 2$.

Proof. The problem is local so we may assume X to be an irreducible analytic subset of the I kind of $\mathbb{R}^n$.

Let $p = \dim X$ and for any $q < p$

${}_qB(X) = \{x \in X \mid$ there exists a component ${}^iX_x$ of X_x such that

${}^iX_x$ is not coherent and $\dim {}^iX_x = q\}$ (see def.3 of § 1).

Clearly we have: $B(X) = \bigcup_{q \leq p} {}_qB(X)$.

The first step is to prove that ${}_pB(X)$ is closed and semianalytic in X. It is obvious that $B(X)$ and ${}_pB(X)$ ($p = \dim X$) are closed.

Let $\tilde{X}^g$ be a completely reduced complexification of X and $P:\hat{X}^g \longrightarrow \tilde{X}^g$ the normalisation of $\tilde{X}^g$.

Let σ, $\hat{\sigma}$ be two antiinvolutions defined on $\tilde{X}^g$, $\hat{X}^g$ such that $X = \{x \in \tilde{X}^g \mid \sigma(x) = x\}$, $P \circ \hat{\sigma} = \sigma \circ P$ (σ and $\hat{\sigma}$ exist (see [5]).

Let $x \in {}_pB(X)$, we may suppose X_x to be irreducible and that $(\tilde{X}^g)_x$ is the complexification of X_x (the problem is local so we may take a neighbourhood of x in X and in the neighbourhood an irreducible component).

We know $(x \in {}_pB(X))$ that there exists $x_n \longrightarrow x$ such that for any realisation $\tilde{V}$ of $\tilde{X}_x = (\tilde{X}^g)_x$ we have: there exists a component $\tilde{V}_{x_m}^{i_m}$ of $\tilde{V}_{x_m}$ such that : $\dim_{\mathbb{R}}(\tilde{V}_{x_m}^{i_m} \cap \mathbb{R}^n) < p$.

For any such $\tilde{V}_{x_m}^{i_m}$ there are two possibilities (see [5]):

α) $P^{-1}(V_{x_m}^{i_m})$ has at least one $\hat{\sigma}$-fixed point.

β) $\hat{\sigma}(P^{-1}(\tilde{V}_{x_m}^{i_m})) = P^{-1}(\tilde{V}_{x_m}^{\sigma(i_m)})$ and $P^{-1}(\tilde{V}_{x_m}^{i_m}) \cap P^{-1}(\tilde{V}_{x_m}^{\sigma(i_m)}) = \emptyset$.

It is known that α) is true iff $\sigma(\tilde{V}_{x_m}^{i_m}) = \tilde{V}_{x_m}^{i_m}$ (see [5]).

We say that $x \in {}_pB(X)$ satisfies α) (or β)) iff there exists $x_{i_m} \longrightarrow x$ such that there exists $\tilde{V}_{x_{i_m}}^{i_m}$ that satisfies α) (or β)).

Let ${}_pB(X)^{\alpha}$ $({}_pB(X)^{\beta})$ be the set of points that satisfy α) (or β)).

We have ${}_pB(X) = {}_pB(X)^{\alpha} \cup {}_pB(X)^{\beta}$.

We shall prove that ${}_pB(X)^{\alpha}$ and ${}_pB(X)^{\beta}$ are semianalytic of codimension at least two $({}_pB(X)^{\alpha}$ has codimension at least three).

By the construction it is clear that: ${}_pB(X)^{\alpha} = P({}_pB(\hat{\tilde{X}} \cap \{\sigma(x)=x\})$.

The set ${}_pB(\hat{X} \cap \{\sigma(x)=x\})$ is semianalytic of codimension at least 3 (see lemma 5), P is a finite morphism, then ${}_pB(X)^{\alpha}$ is semianalytic of codimension at least three (see [9]).

We must prove that ${}_pB(X)^{\beta}$ is semianalytic of codimension at least two.

Let $P^{-1}(X_{p\text{-reg}}) = \bigcup_n A_n$ be the decomposition of $P^{-1}(X_{p\text{-reg}})$ into connected components.

If $x \in A_n$ has the property $\hat{\sigma}(x)=x$ then for any $y \in A_n$ we have $\sigma(y)=y$; in fact we may suppose (by theorem B) that $\tilde{X}^g_x$ is smooth then the set $A'_n = \{z \in A_n \mid \hat{\sigma}(z)=z\}$ is a manifold of dimension p (see [5]) and now it is clear that $A'_n = A_n$.

For any $n \in \mathbb{N}$ let $\sigma(A_n) = A_{\sigma(n)}$; if $n \neq \sigma(n)$ it follows that $A_n \cap A_{\sigma(n)} = \emptyset$.

By construction it is clear that $x \in {}_pB(X)^{\beta}$ iff $P^{-1}(x) \cap \bar{A}_n \cap \bar{A}_{\sigma(n)} \neq \emptyset$ for some $n, n \neq \sigma(n)$.

The set $\bar{A}_n \cap \bar{A}_{\sigma(n)}$ is semianalytic, P is finite so $P(\bar{A}_n \cap \bar{A}_{\sigma(n)})$ is semianalytic, the family $\bar{A}_n$ is locally finite then ${}_pB(X)^{\beta}$ is semianalytic.

Let now $x \in {}_pB(X)^{\beta} - {}_pB(X)^{\alpha}$ then in a neighbourhood of x all components $\tilde{V}^{i_m}_{x_m}$ are of type $\beta)$ and there exists a complex space W_m of dimension $< p$ that contains $\tilde{V}^{i_m}_{x_m} \cap \sigma(\tilde{V}^{i_m}_{x_m})$.

So we may repeat the above argument using $W_{m,(p-1)reg}$ instead of $X_{p\text{-reg}}$ and deduce (see lemma 4) that:

$\dim {}_pB(X)^{\beta} < p-2$.

We have now proved that $\dim {}_pB(X) < p-2$ and ${}_pB(X)$ is semianalytic in X.

We wish now to prove that ${}_{p-1}B(X) \cup {}_pB(X)$ is closed and semianalytic in X. Clearly it is closed.

Let $X_s = X - X_{p\text{-reg}}$, $p = \dim X$, $X(1) \subset X_s$ as in definition 4.

As in the first part of the proof we have ${}_{p-1}B(X) = {}_{p-1}B(X)^{\alpha} \cup$ $\cup\, {}_{p-1}B(X)^{\beta}$ (using a normalisation of a complexification of X_s).

Let $x \in \overline{X(1)}$, if we find $x_n \in X(1)$ such that $x_n \longrightarrow x$ and there exists $\tilde{V}^{i_m}_{x_m}$, $\sigma(\tilde{V}^{i_m}_{x_m}) = \tilde{V}^{i_m}_{x_m}$ with the property $\dim(\tilde{V}^{i_m}_{x_m} \cap \{x \mid \sigma(x) = x\}) < p-1$ then $x \in {}_{p-1}B(X)^{\alpha} \cup {}_pB(X)^{\alpha}$.

If $x \in {}_{p-1}B(X)^{\alpha}$ there exists such a sequence $x_n \longrightarrow x$.

So we have: $x \in {}_{p-1}B(X)^{\alpha} \cup {}_pB(X)^{\alpha}$ iff $P^{-1}(x)$ has a point connection of an handle of the normalisation of X_s that lies in $X(1)$. This proves (see lemma 3) that

${}_{p-1}B(X)^{\alpha} \cup {}_pB(X)^{\alpha}$ is semianalytic in $\overline{X(1)}$ and hence in X.

Let $P: \hat{\tilde{X}}_s \longrightarrow \tilde{X}_s$ be a normalisation of a completely reduced complexification of X_s.

Let $\bigcup_n B_n = P^{-1}(X(1)_{(p-1)\text{reg}})$ where B_n are the connected components.

As in the first part we see that $x \in B_{p-1}(X)^{\beta} \cup {}_pB(X)^{\beta}$ iff there exist two components B_n, $B_{\sigma(n)}$, $n \neq \sigma(n)$, such that:

$$(\bar{B}_n \cap \bar{B}_{\sigma(n)} \cap P^{-1}(x)) \neq \emptyset \ .$$

So we have proved that ${}_pB(X)^\beta \cup {}_{p-1}B(X)^\beta$ is closed and semianalytic of codimension at least two.

In the same way we prove that $B(X) = \bigcup_{i \leq p} {}_iB(X)$ is semianalytic of codimension at least two.

The theorem is proved.

A. Tognoli

Bibliography

[1] W.Fensh "Reel analytische strukturen" Math. Inst. der Univ. Münster Heft 34 (1966).

[2] F.Acquistapace, F.Broglia, A.Tognoli "Sui punti di non coerenza di un insieme analitico reale" To appear.

[3] M.Galbiati "Stratification et ensemble de non-coherence d'un espace analytique réel" To appear.

[4] F.Acquistapace, F.Broglia, A.Tognoli "Sulla normalizzazione di uno spazio analitico reale" To appear.

[5] A.Tognoli "Proprietà globali degli spazi analitici reali" Annali di Matematica 75 pp. 143-218 (1967).

[6] F.Bruhat-H.Whitney "Quelques propriétés fondamentales des ensembles analytiques réel" Com.Math.Helvetici 33 (1959) .

[7] J. Frisch "Points de platitude d'un morphisme d'éspace analytiques complexes" Inventiones Math. 4 pp. 118-138 (1967).

[8] H.Cartan "Variétés analytiques réelles et variétés analytiques complexes" Bull.Soc.Math.France 85 pp. 77-100 (1957).

[9] S.Lojasiewicz "Ensembles semi-analytiques" Inst.des Hautes Études Scient. Bur-sur-Yvette (1965).

A. Tognoli

Pathology and imbedding problems for real analytic spaces

by A. Tognoli(*) (Pisa)

Introduction. H. Cartan has proved (see [2]) that an analytic subset of $\mathbb{R}^n$ may be not coherent and in general it has not global equations. It is also true that the singular set of an analytic subset X of $\mathbb{R}^n$ in general is not contained in a proper analytic subset of X.

Often to construct such examples one takes analytic subset of $\mathbb{R}^n$ union of a non locally finite family of irreducible components.

In this lecture is proved:

1) Any pathology can be realized by irreducible varieties.
2) There exists an analytic subset $V \subset \mathbb{R}^3$ such that : V has only one singular point and V has not global equations.
3) For any $n \in \mathbb{N}$ there exists $V \subset \mathbb{R}^n$ such that: V is a compact analytic subset of dimension two and for any analytic function $f\colon \mathbb{R}^n \longrightarrow \mathbb{R}$ we have $f_{|V} \equiv 0 \Longrightarrow f \equiv 0$.

(*) Lavoro eseguito nell'ambito del G.N.S.A.G.A. del C.N.R..

A. Tognoli

§ 1. Definitions and wellknown facts.

Definition 1. Let (X, O_X) be a ringed space, (X, O_X) is called a real (complex) analytic space iff locally (X, O_X) is isomorphic to a local model $(S, O_{U/\mathfrak{I}})$, where U is open in $\mathbb{R}^n(\mathbb{C}^n)$, O_U = sheaf of germs of analytic functions, $\mathfrak{I}$ is a coherent ideal sheaf of O_U such that S = support of $O_{U/\mathfrak{I}}$.

Definition 2. Let (X, O_X) be a ringed space, (X, O_X) is called a real (complex) analytic variety iff locally (X, O_X) is isomorphic to a local model $(S, O_{U/\mathfrak{I}})$, where U is open in $\mathbb{R}^n(\mathbb{C}^n)$, O_U = the sheaf of germs of analytic functions, $S = \{x \in U \mid f_1(x) = \ldots = f_q(x) = 0,\ f_i \in \Gamma_U(O_U)\}$ $\mathfrak{I}$ = the ideal sheaf of all germs that are zero on S.

Remark 1. It is known (see [1]) that the sheaf O_U is coherent (as O_U-module) and if $S \subset U \subset \mathbb{C}^n$ is locally a locus of zeros of holomorphic functions then the sheaf $\mathfrak{I}_S$ of all germs of analytic functions zero on S is coherent (as O_U module) (see [1]). It follows that a complex analytic variety is a (reduced) analytic space.

A variety is called coherent at the point x if a neighbourhood of x is an analytic space, the variety is called coherent if it is coherent at any point. From the coherence of O_U it follows that a smooth real variety(*) is coherent.

Definition 3. Let (X, O_X) be an analytic space, or an analytic variety, and $Y \subset X$ a closed set. Y is called an analytic set (of X) if, locally, it is the locus of zero of a set of analytic functions.

Remark 2. Usually the supports of analytic spaces and of analytic varieties are assumed paracompact.
In the following we shall also assume that all spaces considered are paracompact.
H. Cartan remarked (see [2]) that a real analytic variety may fail to be a real analytic space. In other words the sheaf $\mathcal{J}_S$ associated to a real analytic set may be not coherent. In the same paper H. Cartan shows that in $\mathbb{R}^3$ there exist analytic sets V for which there does not

(*) Smooth real variety = real analytic manifold.

exist a coherent ideal subsheaf $\mathcal{J}_V$ of $O_{\mathbb{R}^3}$ such that: $V = \text{support } O_{\mathbb{R}^3}/\mathcal{J}_V$.

Definition 4. Let (X,O_X) be a real analytic space and $Y \subset X$ an analytic set. We shall say that Y is non coherent of the I Kind iff the ideal subsheaf $\mathcal{J}_Y \subset O_X$ of all germs vanishing on Y is not coherent (as O_X module), but there exists a coherent ideal subsheaf $\mathcal{J}'_Y \subset O_X$ such that : $Y = \text{support } O_X/\mathcal{J}'_Y$.
If such a $\mathcal{J}'_Y$ does not exist then Y is called non coherent of the II Kind .
In other words Y is non coherent of the I kind iff it is not coherent with the reduced structure , but there exists a non reduced coherent structure.

Remark 3. The category of reduced real analytic spaces is a bad category, for instance the following holds:

i) (X,O_X) may be coherent and reduced but its singular locus not coherent (Ex $\{(x,y,z,w) \in \mathbb{R}^4 | w^2-(z(x^2+y^2)-x^3) = 0\}$) .

ii) Let $X = X_1 \cup X_2$ be a decomposition of a coherent reduced analytic space then X_i is coherent but, in general, $X_1 \cap X_2$ is not coherent (same example as in i)).

The category of real analytic spaces doesn't present such a pathology.

Definition 5. Let (X, O_X) be a real analytic space $(\tilde{X}, O_{\tilde{X}})$ a complex analytic space and $x_o \in X$. We shall say that $(\tilde{X}, O_{\tilde{X}})$ is a complexification of X, at the point x_o, iff $\tilde{X} \supset X$ and $O_{\tilde{X}, \tilde{x}_o} \simeq O_{X, x_o} \otimes \mathbb{C}$ (if $\mathcal{F}$ is a sheaf $\mathcal{F}_x$ is the stalk in the point x). We shall say that $(\tilde{X}, O_{\tilde{X}})$ is a complexification of (X, O_X) iff it is a complexification at every point x of X.

Remark 4. From the definition it is clear that any germ of a real analytic space may be complexified and the germ of the complexification is uniquely determined (in fact it is known (see [4]) that the germ X_x is determined by $O_{X,x}$ and so $O_{X,x} \otimes_{\mathbb{R}} \mathbb{C}$ gives the complexification).
It is easy to check that if X_x is determined, in an open set $U \subset \mathbb{R}^n$, by the equations $f_1 = \dots = f_q = 0$ then $\tilde{X}_x$ is given by $\tilde{f}_1 = \dots = \tilde{f}_q = 0$ where $\tilde{f}_i$ is an extension of f_i to a holomorphic function defined in an open set

$\tilde{U} \subset \mathbb{C}^n$, $\check{U} \supset U$.

The following facts are proved in [2] , [3] , [5] , [6]

α) Let (X, O_X) be a real analytic variety. Then X is coherent in the point $x_0 \in X$ iff the complexification $\tilde{X}_{x_0}$ of the germ X_{x_0} induces the complexification of X_y for all $y \in X$ sufficiently near to x_0 (see [2]).

β) Any real analytic space (possibly non paracompact) has a complexification and a real analytic variety has a complexification iff it is coherent (see [5]).

γ) Let (X, O_X) be a real analytic manifold (i.e. a smooth real analytic space) and $Y \subset X$ an analytic set. Then Y is coherent or non coherent of the I kind iff Y admits global equations(see [3]).

δ) Let (X, O_X) be a normal real analytic variety. Then X is coherent iff $\dim X_x = \dim X$ for any $x \in X$(*) (see [6]).

We shall now discuss some examples given in [2] , [1] .

I) <u>Cartan's umbrella</u>.

Let $V^1 = \{(x,y,z) \in \mathbb{R}^3 \mid \varphi = z(x^2+y^2) - x^3 = 0\}$. It is not difficult to check: that $V^1_{(000)}$ is irreducible and the ideal sheaf associated to $V^1_{(000)}$ is generated by φ .

(*) Not any normal variety is coherent (see [6])

On the other hand near the points $(0,0,z_0)$, $z_0 \neq 0$ the germ V^1 is equal to the germ of the line $x = 0$, $y = 0$. So φ generates the stalk of $\mathcal{J}_{V^1}$ at the origin but not at points of the form $(0,0,z_0)$ and $\mathcal{J}_{V^1}$ is not coherent at the origin.

II) <u>The trascendental Cartan umbrella.</u>

Let $'V^2 = \{(x,y,z) \in \mathbb{R}^3 | \varphi = z(x^2+y^2)-(\exp \frac{1}{z^2-1}) \cdot x^3 = 0$ and $|z| < 1\}$

$''V^2 = \{(x,y,z) \in \mathbb{R}^3 | \quad x = 0,\ y = 0\}$, $V^2 = {'V^2} \cup {''V^2}$.

In [2] H. Cartan proves that if f is an analytic function defined in an open set U of $\mathbb{R}^3$ containing V^2 and $f_{|V^2} = 0$ then $f = 0$. Hence (by γ)) V^2 is a non coherent analytic set of the II kind.

III) <u>An analytic set which is the union of a non locally finite family of irreducible components.</u>

Let ${}^nV^3 = \{(x,y,z) \in \mathbb{R}^3 | (z-n)(x^2+ (y - \frac{1}{n})^2) - x^4 = 0\}$

$'V^3 = \bigcup_n {}^nV^3$, ${}^oV^3 = \{(x,y,z) \in \mathbb{R}^3 | x = 0\}$

$V^3 = {'V^3} \cup {}^oV^3$.

It is not difficult to see that:

$${}^nV^3 \cap \{(x,y,z) | z < n\} = \{(x,y,z) \quad x = 0,\ y = \frac{1}{n} , z < n\}$$

then the family of the parts of ${}^nV^3$ of dimension two is locally finite and V^3 is an analytic set of $\mathbb{R}^3$.

The following is true:

i) any analytic function defined on an open neighbourhood U of V^3 and such that $f_{|V^3} = 0$ is identically zero (in fact if $\tilde{f}$ is an extension of f to a holomorphic function defined on an open set $\tilde{U} \subset \mathbb{C}^3$, $\tilde{U} \supset U$ then $\tilde{f}$ is zero on any complexification of ${}^nV^3$, but such a family of complexifications is not locally finite at some point of U, then $\tilde{f} \equiv 0$).

ii) In V^3 there does not exist an analytic subset Y, of codimension one, such that $V^3 - Y$ is an analytic manifold (Y must contain all the lines $x = 0$, $y = \frac{1}{n}$).

iii) ${}^nV^3$ and ${}^oV^3$ are irreducible components of V^3 but $\bigcup_{n>0} {}^nV^3$ is not an analytic set .

§ 2. Some problems on the pathology of non coherence.

It may be interesting to see how extensive is the pathology illustrated in the previous examples.
For example the following problems are natural:

1°) is it possible to find an analytic set, not coherent of the II kind, that admits only one singular point?

2°) Given $n \in \mathbb{N}$ does there exist a compact analytic set $V \subset \mathbb{R}^n$ of dimension two (any curve is coherent), such that any analytic function $f: \mathbb{R}^n \longrightarrow \mathbb{R}$, vanishing on V vanishes also on $\mathbb{R}^n$?

3°) Let (X, O_X) be a real analytic variety, we shall say that X is strongly irreducible if any analytic set $Y \subset X$ such that $\dim Y = \dim X$ coincides with X .

We may ask if the pathology of example III is possible for strongly irreducible varieties.

We begin with a theorem that says that any pathology is realizable using strongly irreducible varieties.

Definition 1. Let (X, O_X) be a real analytic variety, we shall say that X is pure dimensional of dimension d

iff any closed analytic component of X has dimension d (an analytic component Y of X is an analytic subset of X such that $\mathring{Y} \neq \emptyset$).

Theorem 1. Let (X, O_X) be a real analytic, pure dimensional, variety and X_1 the open set of the smooth points of dimension d(*) $= \dim X$. Then there exist a closed set $F \subset X_1$ and a strongly irreducible variety X' such that: X' has an open set S' isomorphic to $X - F$. Moreover if X is isomorphic to a subspace of $\mathbb{R}^n$ then X' is isomorphic to a subspace of some $\mathbb{R}^m$.

Sketch of the proof.

a) Construction of X' .

We shall suppose that X is countable at infinity and $X = \bigcup_{n=1}^{\infty} X_n$ is the family of irreducible components.

Let

$$C_n = \{(x_1, \ldots, x_d) \in \mathbb{R}^d \mid 1 < (x_1 - 4n)^2 + x_2^2 + \ldots + x_d^2 < 4\}$$

$$L_n' = \{(x_1, \ldots, x_d) \in \mathbb{R}^d \mid 4n - \tfrac{3}{2} < x_1 < 4n - \tfrac{1}{2} , |x_i| < \beta , \beta > 0, \; i = 2, \ldots, d\}$$

(*) Then $x \in X_1$ iff x is smooth and $\dim X_x = \dim X$.

$$L''_n = \{(x_1,\dots,x_d) \in \mathbb{R}^d \mid 4n+\tfrac{1}{2} < x_1 < 4n+\tfrac{3}{2},\ |x_i| < \beta,\ 0 < \beta,\ i = 2,\dots,d\}$$

$A_n = C_n \cup L'_n \cup L''_n$.

On $A = \bigcup_{n=1}^{\infty} A_n$ we consider the following equivalence relation

$$(x_1,\dots,x_d) \overset{\mathcal{R}}{\sim} (y_1,\dots,y_d) \Longleftrightarrow \begin{cases} x_i = y_i & i=1,\dots,d \\ \text{or } x_i = y_i & i=2,\dots,d \\ \text{and } |x_1 - y_1| = 5/2 & \end{cases}$$

Let $V = \bigcup_{n=1} A_{n/\mathcal{R}}$ be the quotient space; V is, in a natural way, a Hausdorff real analytic manifold of dimension d .

For any $n \in \mathbb{N}$ let $P_n \in X_n$ be a smooth point of dimension d of X .

For any $x \in X$ there exists a neighbourhood U_x that contains at most one point of $\bigcup_n P_n$ (the family $\{X_{n,1} = \{x \in X_n \mid x$ is smooth (in X) of dimension $d\}\}$ is locally finite).

Then we can find a locally finite family of open sets D_n of X such that : $\bar{D}_n \cap \bar{D}_m = \emptyset$ if $n \neq m$, and for all n D_n is analytically isomorphic to $D = \{(x_1 \dots x_d) \in \mathbb{R}^d \mid \sum_{y=1}^{d} x_y^2 < 4\}$.

Let $\varphi_n : D_n \longrightarrow D$ be such an isomorphism.

Let :

$D' = \left\{ (x_1 \dots x_d) \in \mathbb{R}^d \mid 1 < \sum_{i=1}^{n} x_i^2 < 4 \right\}$, $D'_n = \varphi_n^{-1}(D')$ $D'' = D - D'$,

$D''_n = \varphi_n^{-1}(D'')$

$F = \bigcup_{n=1}^{\infty} D''_n$ is a closed set contained in X_1 and we may glue $V = A/_{\mathcal{R}}$ to $X-F$ identifying D'_n with C_n. The quotient space $X-F \amalg V/_{\mathcal{R}} = X'$ is the required space.

It is not difficult to verify (see [7]) that :

1) X' has a natural structure of an analytic variety

2) X' is paracompact

3) X' is strongly irreducible (V is a tube that connects all the d-dimensional components of X).

If we do not suppose X countable at infinity it is possible to give a similar construction of X' using a non contable number of D_n (to prove that X' is an analytic variety we use only the fact that $\{D_n\}_n$ is locally finite which is true for general X).

b) <u>The embeddability of</u> X'

We shall now suppose $X \subset \mathbb{R}^n$ and we shall prove that X' (constructed above) can be embedded in some euclidean space $\mathbb{R}^m$.

The proof is contained in two lemmas :

Lemma 1. Let $W' \subset \mathbb{R}^n$, $W'' \subset \mathbb{R}^n$ be two real analytic submanifolds of $\mathbb{R}^n$ and $\varphi : W' \longrightarrow W''$ an analytic isomorphism.

There exist $m \in \mathbb{N}$, $m \leq 2n$, such that, given the embeddings $W' \xrightarrow{i} \mathbb{R}^n \times \mathbb{R}^{m-n}$, $W'' \xrightarrow{j} \mathbb{R}^n \times \mathbb{R}^{m-n}$, $i(x) = (x,0,\dots,0)$, $j(y) = (y,0,\dots,0)$ the isomorphism $j \circ \varphi \circ i^{-1} : i(W') \longrightarrow j(W'')$ can be extended to an isomorphism $\psi : A \longrightarrow B$ where A , B are open neighbourhoods of $i(W')$, $j(W'')$.

Proof. Let T' , T'' be the tangent bundles of W' and W'' , obviously there exists an analytic isomorphism $\alpha : T' \rightarrow T''$.

Let N' , N'' be the normal bundles of $W' \subset \mathbb{R}^n$, $W'' \subset \mathbb{R}^n$.
If E^t is the trivial bundle $W' \times \mathbb{R}^t$ we have (if we identify W' and W'') :

$$\begin{matrix} T' \oplus N' \simeq E^n \\ T'' \oplus N'' \simeq E^n \end{matrix} \Longrightarrow E^n \oplus N' \simeq E^n \oplus N'' \quad (*)$$

(*) We recall that, in the real case, the classication of analytic bundles coincides with the topological one (see [8])

If $V \subset \mathbb{R}^q$ is an analytic submanifold it is known that there exists an open neighbourhood A_V, of V, such that A_V is analytically isomorphic to an open neighbourhood of the zero section of the normal bundle of V in $\mathbb{R}^q$. Using this fact it is clear that ψ' defines an extension $\psi : A \longrightarrow B$ and the lemma is proved.
The theorem is now a consequence of the following

Lemma 2. *Let (X, O_X) be a real analytic variety and suppose $X = X_1 \cup X_2$ where :*

1) *X_1, X_2 are open and $X_1 \cap X_2$ is a manifold*
2) *X_i, $i = 1,2$ is isomorphic to a subvariety of an open set B_i of $\mathbb{R}^n$.*

Then X is isomorphic to an analytic subvariety of an analytic manifold V of dimension $2n$ (and hence $X \subset \mathbb{R}^{4n+1}$).

Proof. Using lemma 1 we know that the identity map $id_{X_1 \cap X_2}$ can be extended to an isomorphism ψ of an open neighbourhood A of $X_1 \cap X_2 \subset B_1 \times \mathbb{R}^n$ onto an open neighbourhood B of $X_1 \cap X_2 \subset B_2 \times \mathbb{R}^n$.
We may now glue, using ψ, an open neighbourhood of

$X_1 \subset B_1 \times \mathbb{R}^n$ to a neighbourhood of $X_2 \subset B_2 \times \mathbb{R}^n$ and construct the manifold V. It is not difficult to prove (see [7]) that V can be constructed so as to be paracompact. So the lemma and the theorem are proved.

Remark 1. Theorem 1 gives a positive answer to problem 3°). We shall now prove that problems 1°) and 2°) have positive answers.

Lemma 3. There exists a strongly irreducible analytic subvariety V of $\mathbb{R}^q$ such that :

1) $\dim V = 2$
2) V is not coherent of the II kind
3) V has only one singular point
4) for any $x \in V$ $\dim V_x = 2$

Proof. Let

$$'V^4 = \left\{(x,y,z) \in \mathbb{R}^3 \mid |z|<1,\ z(x^2+ y^2) - x^3 \exp \frac{1}{z^2-1} = 0\right\}$$

$$''V^4 = \left\{(x,y,z) \in \mathbb{R}^3 \mid x^2 + y^2 + zx = 0\right\}$$

$$V^4 = {}'V^4 \cup {}''V^4 .$$

We remark that V^4 is the union of the trascendental Cartan

umbrella V^2 and of the cone $B = \{x^2 + y^2 + zx = 0\}$.
The singular set of V^2 is the line $x = 0$, $y = 0$ so the singular set of $V^2 \cup B$ is just the origin (B is singular only at the origin).

It is not difficult to prove (see [9]) :

1) V^4 has the origin as singular set
2) V^4 is not coherent of the II kind (obvious because $V^4 \supset V^2$).

If we apply to V^4 theorem 1 the existence of V is proved.

Lemma 4. For any n there exist analytic set $V \subset \mathbb{R}^n$ such that :

1) $\dim V = 2$ and V is compact
2) for any analytic functions $f : \mathbb{R}^n \longrightarrow \mathbb{R}$ we have $f_{|V} \equiv 0 \Longrightarrow f \equiv 0$.

Proof. Let

$$'V_n^5 = \Big\{(x_1,\dots,x_n) \in \mathbb{R}^n \,\Big|\, x_n^2(1-2x_n^2)\Big[(x_1^2+x_n^2-1)^2 + \sum_{j=1}^{n-2} x_j^2\Big] = \\ = \Big[(x_1^2 + x_n^2 - 1)^4 + \sum_{j=1}^{n-2} x_j^4\Big] \exp \frac{1}{x_n^2-1}\Big\} \quad .$$

It is not difficult to prove (see [10]) that $'V^5$ defines an analytic set $V_n^5 \subset \mathbb{R}^n$ such that: V_n^5 is compact, irreducible and any analytic function $f : U_{V_n^5} \longrightarrow \mathbb{R}$ such that $f|_{V_n^5} \equiv 0$ is $\equiv 0$.

Let now p_1 be a regular point of V_n^5 of dimension $n-1$ and $D_1 \ni p_1$ an open set of V_n^5 isomorphic to $\mathbb{R}^{n-1}$. We can realise in D_1 a V_{n-1}^5 and if $p_2 \in V_{n-1}^5$ is a regular point of dimension $n - 2$ we may realise in a neighbourhood of p_2 a $V_{n-2}^5 \cdots$

So we find V_3^5 V_4^5 ... V_n^5 such that for any analytic function $f : \mathbb{R}^n \longrightarrow \mathbb{R}$ with $f|_{V_3^5} = 0$ we have :

$$f|_{V_4^5} \equiv f|_{V_5^5} \cdots \equiv f|_{V_n^5} \equiv 0 .$$

The lemma is now proved.

A. Tognoli

BIBLIOGRAPHY

[1] R. Narasimhan "Introduction to the theory of analytic spaces" Lecture notes in math. - Springer Verlag (1966) .

[2] H. Cartan "Varietes analytiques reelles et varietes analytiques complexes" Bull. Soc. Math. France 85 pp. 77-100 (1957).

[3] F. Bruhat-H. Whitney "Quelques proprietes fondamentales des ensembles analytiques reel" Com. math. Helvetici 33, pp. 132-160 (1959) .

[4] H. Cartan "Seminaire E.N.S. 1960-1961.

[5] A. Tognoli "Elementi di teoria degli spazi analitici reali" To appear on "quaderni Lincei" .

[6] A. Tognoli "Proprietà globali degli spazi analitici reali" Annali di Matematica 75 pp. 143-218 (1967).

[7] F. Acquistapace, F. Broglia, A. Tognoli "Sull'irriducibilità di uno spazio analitico reale" To appear.

[8] A. Tognoli "Sulla classificazione dei fibrati analitici reali" Annali della S.N.S. di Pisa 21

pp. 709-744 (1967).

[9] N. Giannico "Esempio di insieme analitico con singolarità isolata non avente equazioni globali" B.U.M.I. .

[10] F. Acquistapace, F. Broglia, A. Tognoli "Questioni di immergibilità ed equazioni globali" To appear.

Stampa: Editoriale Grafica - Roma - Tel. 5890154